Collins Nature Guide

With love to Dad
on Father's Day.

Pam & Terry

xx

HarperCollins*Publishers*

HarperCollins*Publishers*
77–85 Fulham Palace Road
London W6 8JB

First published 1997

98 00 02 01 99 97
2 4 6 8 10 9 7 5 3 1

ISBN: 0 00 220072 4

Artwork by Sandra Doyle

Colour reproduction by Colourscan, Singapore
Printed and bound by Rotolito Lombarda SpA, Milan, Italy

Introduction

The wide range of habitats found in our gardens makes them extremely rich in terms of the variety of wildlife to be found there. Walls, hedges, trees, lawns, flower beds, vegetable plots, and even sheds and paths all provide board and lodging for plant and animal guests. Some of these visitors – the weeds and pests – can damage or compete with our cultivated plants, but the great majority of garden guests are completely harmless. Some are unquestionably useful in controlling aphids and other pests.

Some of our garden guests are aliens. A number of insect pests, for example, have been introduced accidentally with cultivated plants, while several weeds were originally brought in and grown for food or for garden colour. But most of our garden wildlife consists of native species that have simply moved in from the surrounding areas and found conditions to their liking. Many weeds are characteristic of river banks, cliffs, and other disturbed or unstable places where bare soil is continually being exposed. Freshly-dug gardens are ideal habitats for these plants, which are often brought home as seeds on shoes or clothing. Many insect pests live harmlessly in the wild and become pests only where we grow their food-plants as crops. Large numbers of plants of the same kind growing in a small area allow the insects to increase their numbers dramatically – and that is when they become pests.

The wildlife population of a garden depends on several factors. Northern gardens usually support fewer species than gardens in the south, and urban gardens tend to have a smaller range of wildlife than rural gardens – because it is more difficult for many of the smaller animals to reach them. Air pollution also prevents some species from colonising town gardens, although this is less of a problem now than it was a few decades ago. But the most important factors influencing garden wildlife are the age of the garden and the disposition of the gardener. The older, more mature gardens have a greater diversity of wildlife because the animals and plants have had longer to arrive, and also because there are usually more available habitats – especially trees and shrubs. And this is where the gardener comes in to play, through the kinds of plants that are grown and the way in which they are looked after. A gardener who is always mowing the lawn, trimming the hedges, and hoeing and spraying the borders will have a far poorer and less interesting garden – although perhaps a more productive one – than a gardener who is a little more laid-back and prepared to let nature take its course.

The text and photographs in this book provide information on the identification, distribution, and behaviour of our garden guests and enable the reader to decide whether various creatures are pests or not. Relatively few species cause any real damage and many add colour and interest to the garden. These are well worth encouraging and some ideas for 'wildlife gardening' to attract them appear on the following pages.

Michael Chinery

This book describes about 400 species of plants and animals found in European gardens, including many of the more troublesome weeds and pests. The species are arranged in four main, colour-coded categories.

Colour	Contents	Page
	Vertebrates: animals with backbones. They make up a relatively small part of the animal kingdom – no more than about 45,000 species out of nearly 2 million. Four of the five main groups of vertebrates regularly occur in our gardens; the fifth major group contains the fishes.	Mammals pp.14–23 Birds pp.24–47 Reptiles pp.48–49 Amphibians pp.50–51
	Insects: this group contains more species than all the other animal groups put together. An insect's body is divided into three main regions – the head, thorax and abdomen. The head carries a pair of antennae and the thorax has three pairs of legs. Most insects also have wings.	pp.52–161
	Other invertebrates: spiders have four pairs of legs and their bodies are divided into just two sections. They never have wings. Woodlice are often confused with insects, but they have seven pairs of legs; they are crustaceans. Centipedes and Millipedes are mostly very slender creatures with many pairs of legs. The molluscs include the slugs and snails. They have no legs at all and creep along on a slimy flat foot. Worms also lack legs. All the familiar garden species have clearly ringed bodies.	Spiders pp.162–173 Woodlice pp.174–177 Centipedes pp.178–181 Millipedes pp.182–183 Molluscs pp.184–195 Worms pp.196–197
	Plants: The great majority of plants are flowering plants. The flowers are followed by seeds, which are then scattered in various ways. Many flowering plants invade our gardens. We usually call them weeds.	Mosses pp.198–203 Ferns pp.204–205 Flowering plants pp.206–245 Lichens pp.246–251

The wildlife gardener

It will be obvious from the photographs in this book that many garden guests are extremely attractive and do much to enhance the appearance and interest of a garden. It is certainly worth encouraging them and making them feel at home for our own pleasure, and it is obviously beneficial for the animals themselves. While woods and other natural habitats are being destroyed, the area covered by gardens is increasing all the time. Gardens already cover many thousands of hectares, and they can function as a huge nature reserve if they are managed in the right way.

Bird gardening

There can be few household that do not put at least a few scraps out for the birds, even if they are merely table scraps thrown on to the ground. This elementary form of bird gardening is certainly of benefit to the house sparrows and some of the other common birds, but a little more thought and effort can make the whole business far more interesting for birds and gardeners alike.

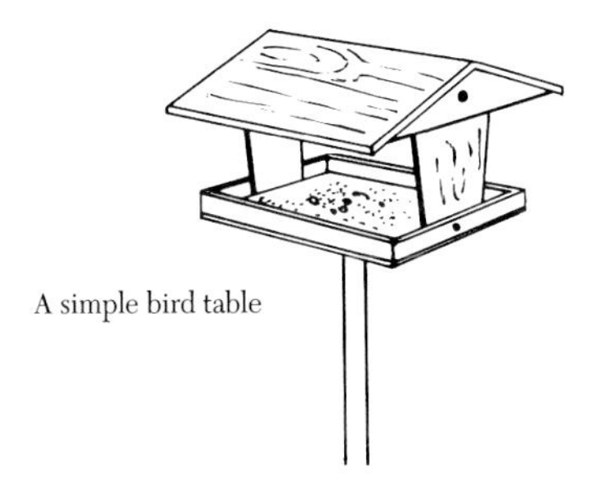

A simple bird table

Although some birds, including the dunnock and blackbird, prefer to feed on the ground, a simple bird table is a much better feeding station for most species. Sited where it can be watched from a window, it can be an endless source of interest. The table should be at least $50cm^2$, to allow for a bit of pushing and shoving, and it should have a shallow rim around most of the perimeter to prevent the food from blowing away as soon as it is put on the table. A few small drainage holes or a couple of small gaps in the rim will allow water to run away and prevent the food from becoming waterlogged. A roof is not necessary, although it will keep of the worst of the rain.

The table can be fixed to a post or a wall or hung from a tree and it must be cat-proof. It should be at least 1.5m above the ground, and any post must be too smooth for a cat to climb. If hung in a tree, the table should be far enough out to prevent cats from reaching it. Ideally, the table should also be squirrel-proof, but this is extremely difficult to arrange. A range of high quality bird tables and other feeding devices can be obtained from the Royal Society for the Protection of Birds.

One of the best ways of feeding kitchen scraps and left-overs is to incorporate them into a pudding with various kinds of seeds and perhaps some cheese and uncooked porridge oats. Mix the ingredients in a bowl and then bind them together with melted dripping or lard. When the mixture has set it can be turned out on to the bird table. A wide variety of birds will find this dish very acceptable. The pudding can also be made in half of a coconut shell with a hook screwed into the end. The shell can then be hung from a branch, or from the bird table and the tits will have a fine time feeding from it.

The pudding can also be pasted into bark crevices, where it may attract woodpeckers, or it can be rammed into holes drilled into small logs, which can then be hung in convenient spots that are visible from the house. Bones can also be hung up for the birds, which will delight the watcher with their antics as they try to remove the last scraps of fat and meat. If they cannot perch on the bones, even starlings and robins make valiant attempts to hover in front of them and make a quick grab.

Finches and other seed-eating birds are most easily attracted with a variety of seeds and dried fruit. Commercially-produced wild bird seed is excellent because it contains a wide range of seeds in various sizes and thus caters for nearly all our seed-eaters. It is worth crushing some of the seed to make it attractive to robins and thrushes and some of the other soft-billed birds, although ready-crushed seed can also be bought as 'songster food'. Bunches of plantain seed heads are worth hanging on the bird table and elsewhere in the garden.

Peanuts are excellent value, especially when presented in wire baskets or other special feeders so that the birds have to work for them. Many such feeders are on the market and some of them claim to be squirrel-proof! Unshelled peanuts threaded on strings or wires attracts various tits and finches and provide equally good entertainment. A fresh coconut cut in half and hung up will also attract plenty of birds, but never offer desiccated coconut, as this swells up in the birds' tiny stomachs and makes life very uncomfortable for them to say the least. And refrain from giving salted nuts or any other salty food because salt is bad for the birds.

Insect-eating birds are happy to peck at nuts, bones, and bacon rind, but there is no better way to attract a robin than by offering mealworms. The larvae of certain beetles, they can be bought from all good pet shops, and they are quite easy to breed in a darkened container filled with bran and dry biscuits – but they must be kept dry. Robins quickly become tame when mealworms are on the menu and readily take them from an outstretched hand. Many other birds, including woodpeckers, also have a liking for the insects.

Regularity is important when feeding birds, for they soon get to know when and where food is put out. They turn up regularly and may become dependent on it, especially in the winter when other food may be difficult to find. It used to be thought that garden birds should be fed only in the winter, but recent research has shown that they benefit by being fed throughout the year – even in the breeding season – although they do not need as much feeding in the summer months when natural food is plentiful. And don't forget that birds also need water. They will appreciate a dish of clean drinking water every day, but especially in frosty conditions and in a dry summer when natural water supplies may be unavailable.

It is also possible to make a garden more attractive to birds by planting a selection of trees and shrubs, although this is feasible only in a large garden. Berry-bearing shrubs are among the most obvious things to grow, for they feed a wide range of birds and the fruit-laden bushes can be very attractive in their own right. Among the most useful bird-attracting shrubs are the various forms of barberry, some of which make good hedges as well as single shrubs, and the numerous forms of cotoneaster – the neat wall-clinging forms as well as the more vigorous 3-dimensional bushes. These shrubs are also very attractive to bees and wasps when they are in flower. *Stranvaesia davidii* is another very good shrub in this respect: honey bees and bumble bees cannot resist its creamy blossoms in May and its waxy red fruits attract numerous birds in the winter. Ivy provides both food and shelter for birds when allowed to clothe old walls and tree trunks, although it can be a nuisance if it gets into the hedge. It flowers in late autumn and several butterflies and other insects appreciate its nectar before going into hibernation.

Hawthorn is a superb multi-purpose tree or shrub. Its rapid growth – reflected in its alternative name of quickthorn – makes it ideal for hedges and its spiny branches provide safe nesting sites for birds: its leaves support many colourful caterpillars: and its abundant fruits, which turn hedgerows maroon in late autumn, feed redwings, fieldfares, and many other birds. Elder also provides an abundance of fruit, which is often removed by blackcaps and other birds before it is properly ripe. But it needs an out-of-the-way corner because it is a rather scruffy tree and its rapid growth can quickly overshadow other plants. It is worth picking some of the fruit in the wild and hanging it on the bird table. It can even be dried for winter use – and, of course, it can be made into excellent wine and jelly. Spindle is another good shrub with colourful fruits, and what about planting a red currant just for the birds. Its fruits ripen early and are eagerly sought out by the blackbirds.

Rowans and birches are well worth planting if space allows. The leaves of both species assume attractive golden shades in the autumn, and the rowan has brilliant orange fruits. The female birch catkins ripen in late summer and autumn and many small birds, including goldfinches, siskins, and various tits, flock to gather the abundant seeds.

Many birds can be persuaded to nest in the garden by providing them with suitable nesting sites, and then it can be fun to watch them bringing in nesting material and food for their youngsters. Trees, shrubs, hedges, and walls clothed with ivy and honeysuckle provide nesting sites for blackbirds, thrushes, and dunnocks. Holes in walls, where bricks or stones have fallen out, make ideal nesting sites for wrens, so it is not a bad idea to leave a few small holes here and there in old walls. The male wren builds several nests and his mate then makes her choice. A good builder may persuade two or three females to occupy his nests.

Several other birds, including the tits, like to nest in some sort of cavity. Tree holes are the usual natural nesting sites of these birds, and most of them are more than happy to move into the artificial holes that we call nestboxes. A wide range of nestboxes can be bought at pet shops and garden centres, but they are easy to make with a minimum of carpentry skills. Small gaps where the wood does not quite meet in the corners provide essential ventilation.

A 'tit-box'

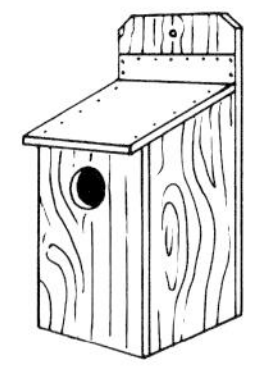

The most popular type of nestbox is the rectangular 'tit-box' (see above), which has a small entrance hole in the front and a hinged, sloping lid to aid cleaning in the autumn (it is not for regular inspection of the nest!). The size of the box is not critical, but the entrance hole must be no more than 29mm across if it is to keep out the house sparrow and encourage the tits. Wrens and treecreepers may also use these boxes.

An open-fronted nest box

Open-fronted boxes are simply rectangular boxes without a front or with just a low wall across the front (see above). Spotted flycatchers, robins, and redstarts are among the birds favouring these boxes. A simple tunnel, made from two pieces of wood, can be tucked under the eaves to accommodate house sparrows. Flycatchers may also use it. One end should be closed if the tunnel is not fitted into a corner. Designs for nestboxes to suit a wide range of birds can be found in *Nestboxes*, which is Guide No. 23 published by the British Trust for Ornithology.

Siting the nestbox is very important. It must be out of the reach of cats and out of the full sun, for a small box can get extremely hot when the sun shines on it. It can be fixed to a wall or a tree and need not be completely vertical, although it is important to ensure that the entrance does not face upwards and allow the rain to get in. Open-fronted boxes are best placed in the shelter of ivy or other climbers. The boxes should be erected in the autumn or winter, so that the birds can examine them and get used to them well before the nesting season. Because most birds are territorial during the breeding season, there is no point in erecting a lot of boxes of the same kind in a small garden. Only one or two are likely to be occupied. It is worth erecting a small landing platform close to the nest entrance. This is not essential for the birds, although some species do like to land and look around before entering the nest, but it does enable the bird-watcher to see what kinds of foods the birds are bringing home for their youngsters.

Homes for bats

Bats are becoming increasingly rare because many of their traditional roosting sites are disappearing. The passion for converting old barns into houses and the widespread use of insecticides in other old buildings have been partly to blame and, although many attics still support colonies of these little mammals, modern houses are less attractive to them than older properties. But the gardener can do a lot to help the bats by providing alternative accommodation on the outside of the house. Bat boxes can be bought from conservation organisations or can be made quite easily at home. They are like tit-boxes but they are much slimmer and there is no opening in the front. The only opening is a narrow slit on the underside of the box.

A bat roost

An even simpler way of providing roosting quarters is to fix a couple of battens vertically to a wall and then to fix a board to them. If the board is placed right up under the eaves the only opening will be a small gap on the underside between the wall and the board. The thickness of the supporting battens should be adjusted so that the gap is no more than about 2cm. This is plenty big enough for the bats to get in and out. The board itself should be at least 2cm thick, in order to provide reasonable insulation, and it should have a rough inner surface. It must not be treated with any preservative, although the outside can be painted to blend in with the house. The size does not matter, but a board 50cm long and 30cm deep will easily accommodate several dozen bats. Bat boxes are best placed on a wall facing west or south-west, to avoid the extremes of heat and cold, or else where they get some protection from nearby trees or other buildings. The animals are notoriously fickle in their choice of roosting sites and it may be several years before new boxes are used: many will probably never be occupied, but it is always worth putting them up. If the bats do take up residence, they may stay there for years, using the boxes for winter hibernation as well as summer roosts and nurseries.

Butterfly gardening

Apart from the injurious cabbage whites (see p.86), the brightly coloured butterflies are usually welcome in the garden. They have been described as mobile flowers, because of the way they dance over the herbaceous borders. About 15 species are regular visitors to British gardens and a few more can be seen in gardens on the continent. These are usually the more mobile species, including the migrants, that search widely for nectar. They are just as likely to occur in town gardens as in rural ones. It is often said that there are fewer butterflies in our gardens today than there were in the past, and this is probably true of most gardens because of the great change in the flowers being grown. Many of today's flowers, including dahlias, chrysanthemums, and various bedding plants, have been bred for colour and size at the expense of nectar and they hold no interest for the nectar-seeking butterflies.

Bringing back the cottage-garden borders with their edgings of aubretia and lavender will bring back the butterflies. Aubretia is particularly attractive to brimstones and tortoiseshells when they wake from hibernation in the spring, while lavender feeds a wide range of butterflies in the summer. Sweet rocket is one of the really traditional cottage-garden plants, sweetly scented and full of butterfly-attracting nectar. Together with honesty, it is also a food-plant of the caterpillar of the dainty orange-tip butterfly. Marjoram is another excellent plant and in rural areas its flowers pull in all sorts of butterflies from the surrounding fields and roadsides. The most famous of the butterfly-attracting plants are undoubtedly the buddleia, aptly named the butterfly bush in many gardening catalogues, and the spectacular ice-plant (*Sedum spectabile*). Coming into flower from June onwards, the various strains of buddleia are alive with peacocks and other colourful butterflies throughout the summer. As the buddleia flowers start to fade in late August, the ice-plant begins to open its plate-sized heads of nectar-rich flowers, and the butterflies waste no time in moving to this rich new source of food. Dozens of small tortoiseshells may gather on a single head, and they are often joined by red admirals, painted ladies, and commas. They all become 'drunk' on the nectar and can be admired and photographed at close range. There are several varieties of ice-plant, but they are not all equally attractive to butterflies. The pale pink varieties are excellent, but the darker 'Autumn Joy' is said to be much less effective.

The plants in the following list, arranged more or less in order of flowering, should attract plenty of butterflies to the garden.

Polyanthus	Phlox
Aubretia	Hyssop
Wallflower	Buddleia
Honesty	Verbena
Sweet Rocket	Marjoram
Red Valerian	Globe Thistle
Bugle	Golden Rod
Mignonette	Aster
Sweet William	Dahlia (single)
Lavender	Michaelmas Daisy
Catmint	Ice-plant

The plants should be massed where possible and in full sunlight. Different varieties often have different flowering times, and by planting two or more varieties you can increase the period of your garden's attractiveness to butterflies. Do not site bird boxes close to your butterfly flowers – certainly not boxes that suit the spotted flycatcher, for this bird will thoroughly enjoy the feast of butterflies.

It is also worth encouraging some of the wild flowers around the garden. Brambles in the hedge, for example, will attract ringlets and gatekeepers in country areas, while a coat of ivy on an old wall will, if allowed to flower, provide a welcome nightcap for butterflies and moths about to enter hibernation in the autumn. It will also provide hibernation sites for brimstones and fodder for the caterpillars of the holly blue butterfly.

Attracting adult butterflies to the garden is quite easy, but permanent increases in the butterfly population can be obtained only by catering for the early stages as well. Unfortunately, apart from the cabbage whites, the caterpillars of most of our garden species feed on what we regard as weeds. Peacock, small tortoiseshell, and red admiral caterpillars, for example, all feed on stinging nettles. Few gardeners have either the room or inclination to set aside areas for these plants, but it is certainly worth leaving patches of long grass under walls and hedges or in orchards to feed the caterpillars of the wall brown and gatekeeper butterflies.

Other possibilities

Attracting bees and wasps may not sound a very good idea to some gardeners, but both groups contain interesting and useful species. Honey bees play an major part in the pollination of fruit trees and other crops, and they are aided in this by the furry bumble bees and a host of solitary bees (see pp.154–57). Most bumble bees nest on or under the ground in rough, grassy places, especially in sunny hedgebanks. Natural nesting sites are fairly plentiful, but queens can sometimes be persuaded to nest in up-turned flower pots buried almost completely in the ground. The success rate can be improved by packing shredded grass into the pot, and even more so if some old mouse bedding – scrounged from a pet shop perhaps – can be added to it .

The flower pot (above left) has been buried in a bank and contains mouse bedding. A piece of hosepipe forms the entrance to the nest. A nest that is placed in level ground (above right) is kept dry using a piece of slate balanced on two stones.

Many of the solitary bees nest in small burrows in the ground or in tunnels in wood and masonry. Although they often make their own tunnels, they happily move in to ready-made cavities if we provide them. Holes of varying diameters drilled into logs and tree stumps make excellent bee dwellings, and the insects will also nest in short lengths of cane or hogweed stems plugged at one end and glued under window sills. Some of the smaller species will even nest in drinking straws, especially if these are fixed to walls or tree trunks in small bundles. Ventilation bricks scattered around the garden or built into a small wall are commonly occupied by mason bees (see p.154). If the bees do move in, it is interesting to watch them bring in mud and other materials to make their cells, and later they can be seen bringing in loads of pollen – indicating that our flowers are being pollinated as well.

Solitary wasps (see p.148) often nest in the holes and tunnels provided for the bees, and they can be seen stocking their nests with a wide range of insects – many of which are likely to be garden pests.

A wildlife pond

Even a small pond can add a great deal of interest to a garden and, with the loss of so many farm and village ponds and the pollution of many of those that do remain, garden ponds are playing an important role in the conservation of frogs and other aquatic wildlife. But it is not only the aquatic creatures that enjoy the garden pond: swallows and wagtails will gratefully accept the small flies that emerge from the water, and many other birds may visit the pond to bathe and drink. Hedgehogs may also come for an evening drink.

There are several ways of making a pond, but the simplest method is undoubtedly to buy a fibreglass pond and to drop it into a hole dug to the correct size and shape. The other recommended method is to use a flexible butyl liner. These liners are expensive, but they will fit any desired shape. Unfortunately, there is still no substitute for the hard work of digging the hole. The sides of the hole should slope gently and there should be a number of shelves (or a continuous shelf) on which marginal plants can be grown. The bottom should be at least 50cm deep, and preferably more to ensure that the pond does not freeze solid in a hard winter. One end can be left very shallow if desired, and planted with marsh marigolds and other mud-loving plants. Ponds should always be sited in fairly open spots where they get plenty of sunshine.

The size of the liner required should be calculated by measuring the maximum length and width of the pond and adding twice the maximum depth to each dimension. Sharp stones must be removed before the liner is put in place, and the hole should ideally be lined with soft sand and old newspapers. An old carpet is even better. The liner moulds itself to the shape of the hole when water is added, and when the pond is full the edges can be concealed under stones or turf. The edges should slope away from the pond so that the rain does not wash dirt and other debris into the water. A large log wedged across one end of the pond will enable frogs and newts to get in and out easily and will give them some good sunbathing spots. Alternatively, a large boulder can be placed in the pond.

Tapwater is usually the only water available for filling the pond and, contrary to popular opinion, it can safely be used for this purpose. It does not actually harm pond life, although it tends to be rich in minerals and it may turn green when the algae multiply in the summer. But this is not a serious problem and the algae gradually dwindle as they use up the minerals. Rainwater or water from another pond should always be used for topping up the pond in summer. It is also a good idea to 'seed' a new pond with water from an existing pond or water butt in order to introduce the microscopic life on which most other pond life depends.

Water milfoil, hornwort, and Canadian waterweed (*Elodea*) are among the best plants for the pond. Simply dropped into the water, they will quickly form dense clumps and give off all the oxygen needed to keep the pond and its animal life in good shape. Purple loosestrife, arrowhead, irises, and other emergent plants can be planted in plastic baskets placed on the ledges around the margins of the pond, and water lilies can be planted in the deepest part of the pond if desired. All these plants can be bought from garden centres or scrounged from friends. Most people with ponds are only too pleased to pass on their excess vegetation.

Animal life can also be scrounged from neighbouring ponds, but most animals will quickly find new ponds by themselves. Newts and frogs are often among the first arrivals and they may breed in the first year. Toads are less likely to breed to start with because they like their traditional breeding ponds. Adding a few toad tadpoles from elsewhere is the best way to establish a toad population. Dragonflies zoom over the pond in search of small flies, and may well breed there in company with water bugs and water beetles. Water fleas and other tiny creatures may arrive in mud clinging to the feet of birds or other animals, and the eggs of larger animals such as water snails may also arrive by this route. Before long the new pond will support a thriving community.

Note: Unless otherwise stated, the measurements given for animals are the average body lengths, without the legs, although the length of the forewing is given for butterflies and moths because this is easier to assess. The measurements given for birds are total lengths from the tip of the beak to the tip of the tail. The figures given for plants are the normal maximum heights. Species listed under *Similar species:* include only those likely to be found in gardens. Species found in other habitats are not included.

1 Hedgehog

Erinaceus europaeus

Description: Head and body up to 27cm: tail 1.5–3cm. The coating of about 6000 spines makes this animal quite unmistakable.

Food: Worms, slugs, beetles, and many other invertebrates, together with fallen fruit and the occasional young bird or rodent.

Habits and signs: Almost entirely nocturnal, with much snorting and snuffling as it searches for food. Squeals loudly when alarmed and also during courtship. Droppings, up to 4cm long and often found on lawns, are black and tarry and contain numerous insect remains. Hibernates in coldest months.

Habitat and range: Woods, hedgerows, parks, and gardens. Throughout Western Europe except northern Scandinavia.

2 Mole

Talpa europaea

Description: Head and body up to 16cm: tail up to 4cm. Easily identified by its velvety black fur and spade-like front feet.

Food: Mainly earthworms, which make up over 90 per cent of the diet in winter. Also eats slugs, millipedes, and insect larvae in the soil.

Habits and signs: Spends almost all its life tunnelling through the soil with its enormous front legs, pushing up tell-tale mole-hills of excavated soil at intervals. Tunnels may be as much as 100cm below ground, but are sometimes just under the surface and visible as snaking ridges. Occasionally forages above ground. Young moles also wander over the surface in summer when they leave home and look for new territories.

Habitat and range: Woods and grassland, including orchards and large lawns. Generally absent from regularly cultivated land and from very sandy or stony soils, which do not hold enough earthworms. Most of Europe except Ireland, Scandinavia, and the far south.

3 Common Shrew

Sorex araneus

Description: Head and body up to 8.5cm: tail up to 5.5cm. Dark brown to blackish on the back, brown on the flanks, and dirty white underneath. Tail hairless when adult. The long, pointed snout, typical of all shrews, readily distinguishes the animal from mice and voles.

Food: Earthworms, slugs, beetles, and many other invertebrates.

Habits and signs: Active night and day, but much of its activity is just below ground or in runways through leaf litter. Presence often indicated by high-pitched squeaking. Often caught and brought in, but rarely eaten by domestic cats.

Habitat and range: Dense grassland (especially roadside verges), hedgerows, woods, and orchards. Common in gardens surrounded by hedges. Most of Europe, but absent from Ireland, Iberia, and most of France.

Similar species: Millet's Shrew of France and N. Spain is externally indistinguishable. Pygmy Shrew lacks brown flanks. Several other species, all with hairy tails, occur in gardens on the continent.

1

2

3

Bats

Bats are the only flying mammals. Their wings consist of thin skin supported on the arms and very long fingers and extending back to enclose the hind legs. Mainly nocturnal, the European species all feed on insects, especially midges and other small flies, which are usually caught on the wing in full flight. The bats spend the daylight hours asleep in hollow trees, caves, roof spaces, and similar places, with their wings wrapped tightly around their bodies. All European bats hibernate through the winter. Several bat species hunt over our gardens. Most of them are very similar in flight and, although their sizes vary, expert knowledge is necessary to identify them. All bats are legally protected and roosts should never be disturbed. The local Wildlife Trust can help in situations where bat roosts may cause problems.

1 Brown Long-eared Bat

Plecotus auritus

Description: Head and body 3.5–5.5cm: wingspan 24–28cm. Ears brown and very long – over half the body length. Fur long and fluffy, light brown on the back and yellowish brown to white underneath. Wings brown.
Habits and signs: Entirely nocturnal, with a rather slow, fluttery flight, usually keeping low over the vegetation. Moths and other insects are usually plucked from leaves but sometimes taken in flight: often taken to a perch to be eaten. Feeding sites are often marked by piles of moth wings. Hibernation is usually from late October to March.
Habitat and range: Open woodland, parks, orchards, and mature gardens. Roosts in hollow trees and buildings and regularly uses bat boxes: one of the commonest bats in roof spaces. All Europe except the far north and south. The second most common bat in the British Isles after Pipistrelle.
Similar species: The Grey Long-eared Bat is very similar in habits and appearance but has greyish fur. It has a more southerly distribution, including southern England.

2 Pipistrelle

Pipistrellus pipistrellus

Description: Head and body 3.5–5cm: wingspan 18–24cm. Ears and wings blackish brown. Fur ranges from orange to dark brown on back and from yellow to greyish brown underneath, but the basal part of each hair is always dark. The smallest and one of the commonest European bats.
Habits and signs: Entirely nocturnal, usually emerging from roost at dusk and flitting silently through the air with a rapid and rather jerky flight, although some people can hear some of their high-pitched calls. The bats commonly roost in house roofs. Their droppings are often mistaken for mouse droppings, but they are very crumbly and are easily identified by the insect remains in them. The bats usually hibernate from November to March, but may fly on mild nights in the winter.
Habitat and range: Woodland margins and most lightly wooded areas, but especially common in and around towns and villages: often seen feeding on moths and other insects attracted to lights. All Europe except the far north.

1

2

1 Grey Squirrel

Sciurus carolinensis

Description: Head and body 25–30cm: tail up to 25cm. Back and tail always grey, although flanks may have a reddish tinge.
Food: Nuts, especially acorns and hazel nuts, and other fruits, buds, and young twigs: strips bark to eat nutritious tissues below. Also eats insects, birds' eggs, and nestlings, and regularly 'steals' from bird tables. A serious forest pest, also damaging garden and orchard trees.
Habits and signs: Diurnal, keeping mainly to the trees. Neatly split hazel nut shells and roughly chewed cones litter feeding sites.
Habitat and range: Mixed woodland, parks, orchards, and gardens. A North American animal, introduced to the British Isles late in 19th century and now common all over England and Wales, where it has largely replaced the native Red Squirrel: less common in Scotland and Ireland.
Similar species: Edible Dormouse (below) is smaller.

2 Red Squirrel

Sciurus vulgaris

Description: Head and body 18–24cm: tail up to 20cm. The fur is bright, reddish brown in summer, but dark brown in winter, often with a grey tinge on the body and tail. Ear tufts are prominent in winter.
Food: Conifer seeds, acorns and other nuts, fungi, and some insects.
Habits and signs: Diurnal. Feeding signs are like those of Grey Squirrel.
Habitat and range: Woodland, parks, and gardens. Throughout Europe, but confined to coniferous forest in the British Isles.

3 Garden Dormouse

Eliomys quercinus

Description: Head and body 10–17cm: tail up to 15cm. Yellowish brown or greyish above and white below: a prominent black patch from in front of the eye to behind the ear. Tail ends in a bushy black and white tuft.
Food: Nuts and other fruits, buds, and assorted invertebrates: also eats birds' eggs and nestlings and young mice and voles.
Habits and signs: Largely nocturnal, but sometimes seen at dawn. Climbs well, but often feeds on the ground. Hibernates September to March.
Habitat and range: Woodland, scrub, orchards, and gardens, often nesting in outbuildings. Southern and central Europe, but not the British Isles.

4 Edible Dormouse

Glis glis

Description: Head and body 13–20cm: tail up to 15cm and very bushy. Grey, often strongly tinged with brown, with a dark ring round the eye.
Food: Nuts and other fruits (especially apples), fungi, various small animals.
Habits and signs: Nocturnal, living in small groups and often very noisy, especially when living in roof spaces. Hibernates October to March.
Habitat and range: Deciduous woodland, orchards, gardens; usually keeps well above ground. Southern and central Europe: introduced to England and well established in the Chiltern area, where often called glis-glis.
Similar species: Grey Squirrel (above) is bigger, with no dark eye ring.

1

2

3/4

1 Bank Vole

Clethrionomys glareolus

Description: Head and body 9–12cm: tail up to 7cm. Snout quite blunt. Fur on back distinctly reddish brown, often with greyish flanks: underside dirty white.

Food: Fruits and seeds, leaves, moss, fungi, and small invertebrates.

Habits and signs: Active both day and night. Climbs well and can often be seen eating blackberries, elderberries, and other fruit in trees and bushes. Hazelnut shells with neat round holes and no obvious tooth marks indicate attack by bank voles. Corpses often brought in by cats.

Habitat and range: Woods, hedgerows, and scrubby places, including shrubberies and rural gardens. All Europe except far north and south.

Similar species: Grey-sided Vole of Scandinavia is largely grey, with just a band of reddish fur along the back. Few other voles are likely to be seen in gardens, although the Field Vole may occur in rural gardens surrounded by grassy habitats. It is dull brown, with a very short tail.

2 Wood Mouse

Apodemus sylvaticus

Description: Head and body 8–11cm: tail 7–11cm. The fur is dark brown on the back, yellowish on the flanks, and off-white underneath. The snout is pointed and the ears are very large.

Food: Eats whatever food is available, including fungi, worms, and insects – especially caterpillars. Fruits and seeds, often taken from the bushes, are a major food source in autumn and winter.

Habits and signs: Largely nocturnal, climbing well in bushes and small trees. Seeds and small fruits, such as haws, are often stored under stones or in log-piles, and sometimes in old birds' nests. Hazelnuts attacked by wood mice have circular holes surrounded by tooth marks.

Habitat and range: Woods, hedgerows, scrub, and cultivated land of all kinds. Throughout Europe except the far north.

Similar species: Yellow-necked Mouse has a large yellow patch – often a complete collar – just in front of front legs.

3 House Mouse

Mus musculus

Description: Head and body 7–10cm: tail up to 10cm, about same length as body. The fur is dull greyish brown above, gradually grading to a slightly lighter grey below. A strong musky smell.

Food: Omnivorous, with a preference for grain and other seeds.

Habits and signs: Largely nocturnal. Climbs well, especially in buildings. Nibbles domestic fabrics and uses them as bedding. Sausage-shaped droppings, up to 6mm long, are often concentrated in certain spots. They differ from bat droppings in not crumbling when squashed.

Habitat and range: Most common around human habitation, especially on farms and rubbish dumps and in grain stores. Often nests in sheds and houses, frequently in roof spaces and under floor-boards. Worldwide.

Similar species: Wood Mouse (above) is browner, with much larger ears.

1

2

3

1 Brown Rat

Rattus norvegicus

Description: Head and body 20–23cm: tail 17–23cm. Fur generally brownish grey above and pale grey below, but ranges from pale brown to black. Ears rather short and hairy.

Food: Omnivorous, with a bias towards grain and cereal products: eats meat in warehouses and other stores. Gnaws many inedible materials.

Habits and signs: Largely nocturnal, resting in shallow tunnels by day. The animals make conspicuous trails through vegetation, especially at the base of walls and hedges, and leave greasy patches wherever they regularly brush against walls and other objects. The spindle-shaped droppings are up to 2cm long.

Habitat and range: Mainly associated with human settlements, especially around farms, sewers, and rubbish dumps. A recent survey suggested that few gardens are without the brown rat and that people are rarely more than a few yards from one of these animals. Worldwide.

Similar species: Black Rat, now almost extinct in the British Isles but common in south and central Europe, has larger hairless ears.

2 Red Fox

Vulpes vulpes

Description: Head and body 60–90cm: tail 30–50cm and very bushy. An unmistakable animal.

Food: A wide range of animals, including rabbits, rats and other rodents, birds (including domestic chickens if it can get at them), beetles, and worms. Carrion is readily taken and the fox is a notorious scavenger at dustbins. Fallen fruit is eaten in autumn.

Habits and signs: Largely nocturnal, but often seen by day in undisturbed habitats. Leaves a pungent odour wherever it goes. Communicates by a wide range of barks and shrieks. Droppings, often left on prominent stones, are long and dark and, unlike those of domestic dogs, commonly have curly ends caused by the fur in them.

Habitat and range: Almost any habitat, including towns and cities. Urban foxes hunt in parks and large gardens and often breed in cemeteries and along railway lines. Throughout the northern hemisphere.

3 Beech Marten

Martes foina

Description: Head and body 40–50cm: bushy tail 25–30cm. The fur is dark brown, with a divided white bib.

Food: Mainly rodents, but takes some birds and insects, and also fruit. Carrion and kitchen waste are readily eaten.

Habits and signs: Mainly nocturnal, but may be active by day in undisturbed woodland. Usually solitary. Droppings, up to 10cm long and often somewhat coiled, are left at specific sites, such as barn ledges.

Habitat and range: Woods and rocky places: common in many towns and villages, often living in barns and roof spaces – where they can be rather noisy at night. All Europe except British Isles and Scandinavia.

1

2

3

1 Kestrel

Falco tinnunculus

Description: 35cm. Male's grey head and tail and spotted chestnut back are very characteristic. Female has a brown head and tail as well, with conspicuous dark bands on the tail. The underparts are pale with dark streaks in both sexes. Wings are very pointed.
Voice: A shrill 'kee-kee-kee'.
Feeding habits: Feeds mainly on voles and other small mammals, spotted while hovering about 15m above the ground: also takes beetles and other insects and is not averse to snatching birds from the bird table.
Nest: A bare hollow, usually on a rock ledge or building but sometimes in a tree: occasionally adopting the old nest of another bird.
Habitat and range: Mainly open country, often hovering over roadsides, but not uncommon in towns and villages: often breeds on church towers and other tall buildings. Most often visits gardens in winter.
Similar species: Sparrowhawk, which also snatches birds from bird tables, is often mistaken for kestrel, but wings are grey or brown and less pointed: underparts are distinctly barred.

2 Collared Dove

Streptopelia decaocto

Description 32cm. Greyish brown back and wings, with dark wing-tips and a conspicuous black collar at the back of the neck. Underparts pale grey, often with a pink tinge.
Voice: A monotonous, repetitive 'coo-cooo-cu', usually delivered from trees or roof-tops. The accent is on the second note and the third note is sometimes very short or dropped altogether. The call can then be mistaken for that of a cuckoo, but the cuckoo's call is more musical and has the accent on the first note.
Feeding habits: Feeds mainly on grain and other seeds on the ground.
Nest: A thin platform of twigs in trees or on buildings.
Habitat and range: Mainly towns and villages: rarely far from human habitation. Most of Europe, after a remarkable expansion from the Balkans since 1930.

3 Hoopoe

Upupa epops

Description: 28cm. Pinkish brown body and boldly banded black and white wings. The crest on its head is raised like a fan for a few seconds when the bird alights, but otherwise only when it is excited. Unmistakable.
Voice: A soft and rather haunting 'hoo-poo-poo', which gives the bird its name. Also a mewing sound.
Feeding habits: Feeds mainly on ground-living insects, often stabbing its long beak into turf to grab them.
Nest: An unlined hole, usually in a tree but sometimes in a wall.
Habitat and range: Lightly wooded country, including parks, orchards and large gardens: often feeds on lawns. A summer visitor to southern and central Europe, but rarely seen in Britain.

1

2

3

1 Lesser Spotted Woodpecker

Dendrocopos minor

Description: 15cm: the smallest European woodpecker. Conspicuous black and white bars on back and wings: no red under the rump. The male crown is all red; the female crown is dirty white with a black patch at the rear.
Voice: A high-pitched, repetitive 'pee-pee-pee'. The bird also drums in the spring by hammering on dead branches with its beak. Each burst or roll lasts for about two seconds, during which time the bird hits the branch up to 30 times. Drumming is most often heard in woodland.
Feeding habits: Mainly insects, taken from twigs and bark crevices, but also takes fat and nuts from garden bird-feeders.
Nest: A simple, unlined cavity, usually in a dead tree.
Habitat and range: Woods, parks, and old orchards: mainly a winter visitor to gardens. Most of Europe, but absent from Ireland and much of Scotland.

2 Great Spotted Woodpecker

Dendrocopos major

Description: 23cm. Buff forehead, with black back and conspicuous white wing patch. A black bar separates the white cheek from a white patch on the neck. Bright red patch under the rump. Female lacks red nape. Juvenile has red crown.
Voice: A harsh 'chick', often repeated several times in quick succession, as well as various chattering and churring sounds. It drums more loudly than the lesser spotted woodpecker, but each burst lasts for no more than a second.
Feeding habits: Feeds mainly on wood-boring insects taken from tree trunks, but also eats nuts and seeds: often wedges cones into bark crevices and then digs out the seeds. Regularly takes fat and nuts from bird tables.
Nest: An unlined cavity in a tree. Sometimes enlarges holes in tit-boxes and nests there.
Habitat and range: Woods, parks, orchards, and gardens. All Europe except Ireland.

3 Green Woodpecker

Picus viridis

Description: 32cm. Both sexes are easily identified by the green body and red crown. Yellow rump clearly visible in flight which, as in most woodpeckers, is distinctly bouncy or undulating.
Voice: A loud, chuckling 'plew-plew-plew', heard in spring, gives the bird its alternative name 'yaffle'. At other times it produces a cackling yelp. Rarely drums.
Feeding habits: Takes insects from trees, but also eats a lot of ants on the ground and often digs into lawns to find them.
Habitat and range: Woods, parks, and gardens all over Europe except Ireland and the far north.

1

2

3

1 House Martin

Delichon urbica

Description: 12cm. Black upperparts, apart from a white rump, and pure white underparts. Short, forked tail and very short beak. Usually seen streaking through the air or perched on wires and aerials.
Voice: A soft twittering sound, uttered in flight or while perched.
Feeding habits: Entirely insectivorous, collecting flies and other small insects in flight.
Nest: Cup-shaped, made with mud and various plant fibres and fixed under the eaves of buildings or other similar overhangs: a narrow entrance at the top. Birds are often seen gathering mud on the ground.
Habitat and range: Most common in and around towns and villages, but also nests on cliffs and quarry walls and on bridges. A summer visitor to all parts of Europe.
Similar species: Swallow is similar in flight, but adult has long tail streamers.

2 Swallow

Hirundo rustica

Description: 20cm. Glossy blue-black above, with red forehead and throat. Underparts white with a dark blue band just below the throat. Long tail streamers are present only in the adult. Usually seen streaking through the air or sitting on wires and aerials.
Voice: A high-pitched twitter uttered in flight and when perched.
Feeding habits: Collects flies and other small insects in mid-air.
Nest: A shallow cup made of mud and plant fragments and lined with feathers: mainly in barns and other buildings and under bridges. Nest is always built on a small ledge and attached to a vertical face as well.
Habitat and range: Mainly in and around towns and villages. A summer visitor to all parts of Europe.
Similar species: House Martin is similar in flight, but has a shorter tail.

3 Pied Wagtail

Motacilla alba

Description: 18cm. Easily recognised by its long tail and black, grey, and white coloration. British specimens have a black back and rump: continental birds, known as white wagtails, are grey. The throat becomes white in winter. Most often seen running jerkily over lawns and paths, stopping frequently and wagging its tail vigorously.
Voice: A high-pitched twitter based on a mixture of 'chick' and 'chizzick'.
Feeding habits: Catches insects on or close to the ground, and sometimes flying over water. Also takes crumbs under the bird table.
Nest: A cup made from plant materials and lined with feathers: built in holes in walls and other cavities. May use open-fronted nestboxes.
Habitat and range: All kinds of open country, including gardens with large lawns: usually near water. Resident in southern Europe and British Isles: summer visitor elsewhere.

1

2

3

1

Spotted Flycatcher

Muscicapa striata

Description: 14cm. Dull brown back and wings: forehead paler with dark streaks. Underparts dirty white with dark streaks on breast.
Voice: A shrill, squeaky 'tzee-tzee-tzee'.
Feeding habits: Entirely insectivorous, almost always taking prey on the wing. The bird adopts a regular perch from which it darts out to snap up passing insects, and usually returns to the perch to eat them.
Nest: A cup of grass and other vegetable material, lined with hair and feathers. Usually built in holes or on well-protected ledges: not uncommon on house walls clothed with ivy or other creepers. Readily uses open-fronted nestboxes.
Habitat and range: Lightly wooded areas, including orchards, parks, and gardens: quite common in towns. A summer visitor to all parts of Europe.
Similar species: Dunnock lacks streaked breast.

2

Treecreeper

Certhia familiaris

Description: 13cm. Streaky brown above, with a pure white underside and a long curved bill. Usually seen creeping up tree trunks, often taking a spiral route as they explore the crevices.
Voice: A high-pitched 'tsee-tsee-tsee', often punctuated by shorter notes.
Feeding habits: Insectivorous, finding most of its food in bark crevices. Also takes fat and other food presented in crevices.
Nest: An untidy cup of twigs, moss, and grass, lined with feathers and wool: usually built behind loose bark or amongst ivy on trees and walls.
Habitat and range: Woods, orchards, town parks, and gardens with mature trees: rarely far from trees. Most often seen in gardens in winter. Most of Europe except the Arctic and the south-west.
Similar species: Can be mistaken for a sparrow from a distance, but white underparts and curved bill readily distinguish it.

3

Wren

Troglodytes troglodytes

Description: 9cm. A plump, rounded bird with distinctively barred rich brown wings and back and a short, often cocked tail. Often mistaken for a mouse as it scampers through the undergrowth.
Voice: A strident 'tic-tic-tic', repeated in quick succession when alarmed, and a clear warbling song: surprisingly loud for such a small bird.
Feeding habits: Eats insects, spiders, and other small invertebrates, and occasionally takes crumbs on and around the bird table.
Nest: A ball of moss, leaves, and other plant material with an entrance at one side: usually built in a hole in a wall, bank, or tree trunk. The male generally builds several nests, but not all are used: the female chooses one and lines it with feathers before laying her eggs. Sometimes nests in tit-boxes, and roosts in them in winter.
Habitat and range: Hedgerows and other places with plenty of dense cover. Most of Europe, but only a summer visitor to northern Scandinavia.

1

2

3

1 Blue Tit

Parus caeruleus

Description: 12cm. The bright blue crown and yellow breast make this a very easy bird to recognise.
Voice: The bird has a wide range of calls and songs, most of which are variations on a theme of 'tsee-tsee-tsit' or 'tsee-tsee tsu-tsu'.
Feeding habits: Omnivorous, but with a strong bias towards caterpillars and other insect food: buds, small seeds, and fruit contribute to the diet in the wild, and the bird will take almost anything from the bird table. Hacks through milk-bottle tops to get at the cream.
Nest: Builds in a hole, usually in a tree or a wall, and readily accepts a nestbox with a small hole. The nest, built by the female, is made mainly from moss and dead leaves and is lined with hair and feathers.
Habitat and range: Woods, hedgerows, gardens, parks, and almost anywhere else with scattered trees and shrubs. A recent study of garden birds by the British Trust for Ornithology revealed that blue tits were feeding in every garden surveyed during the winter. Throughout Europe apart from northern Scandinavia.
Similar species: Great Tit has a black crown.

2 Great Tit

Parus major

Description: 14cm. Black crown and bib, bluish-grey wings, and bright yellow underparts with a central black line extending down from the bib.
Voice: A very wide range of calls and songs, including 'tsee-tsee-tsee, pink-pink-pink', and 'pitwee-pitwee-pitwee'. The latter song has a bell-like tone, has been likened to the sound of pumping up a bicycle tyre.
Feeding habits: Omnivorous: may damage buds in spring, but more than compensates for this by destroying large numbers of injurious caterpillars. Takes almost anything from the bird table, and also steals cream from milk bottles.
Nest: Builds in a hole like Blue Tit and regularly uses nestboxes.
Habitat and range: Woods, gardens, parks, orchards, and anywhere else with scattered trees and shrubs. Throughout Europe.
Similar species: Blue Tit has bright blue crown.

3 Coal Tit

Parus ater

Description: 12cm. Black crown and throat, with a white patch on the nape. The wings are sooty grey and the underparts are dirty white.
Voice: A wide range of calls, many of them similar to those of the Great Tit but higher-pitched. The most common song is a bell-like 'teacher-teacher-teacher'.
Feeding habits: Searches for insects and spiders in trees and for seeds on the ground. Less frequent at bird tables than blue and great tits.
Nest: In a hole and very like that of Blue Tit. Sometimes uses nestboxes.
Habitat and range: Woods, parks, and gardens, especially with conifers. Mainly a winter visitor to gardens. All Europe except the far north.

1

2

3

1

Blackcap

Sylvia atricapilla

Description: 14cm. Wings and back brownish grey. Crown sooty black in male, rust-red in female, pictured here. Underparts pale grey, tinged with buff in female. There is no black bib.
Voice: Commonest call is a sharp 'tac-tac-tac'. The song, considered to be among the best in the British Isles, is a very musical warble, usually starting quietly and ending with a flourish after about five seconds.
Feeding habits: Insects, spiders, and small snails are the main foods in spring. Later in the year the birds attack cherries, pears, and many other fruits. They take nuts and other foods from bird tables in winter.
Nest: A neat cup made of dry grass and roots and lined with fine grass and hair. Built in dense bushes and hedgerows.
Habitat and range: Woodland and scrub, town parks, and large gardens. Resident all year in southern Europe and on the Atlantic seaboard as far north as southern England. A summer visitor to most other parts.

2

Black Redstart

Phoenicurus ochruros

Description: 14cm. Male is sooty black, with a white wing patch and rust-red tail: female is dull brown with rust-red tail. A rather restless bird, rarely perching for long and continually flicking its tail to display the red feathers.
Voice: Main call is a rapid 'tic-tic-tic'. The song is a squeaky warble.
Feeding habits: Largely insectivorous: plucks spiders and caterpillars from leaves and often darts out from a perch to snatch insects in mid-air. Also eats berries in autumn.
Nest: A cup, made of dry grass and other plant material and lined with hair and feathers. Built by female in holes in rocks and buildings: sometimes on ledges in barns.
Habitat and range: Cliffs and other rocky areas, including ruined buildings: also in towns and villages on the continent. Known as 'house redstart' in Germany because of association with houses. Resident in the south and on the Atlantic seaboard as far as southern England: a summer visitor elsewhere, but not Scotland, Ireland, or most of Scandinavia.

3

Redstart

Phoenicurus phoenicurus

Description: 14cm. Breeding male has grey back and wings, an extensive black throat, a white forehead, and a rust-red tail and underparts. Female is dull brown with buff underparts and a rust-red tail. Continually flicks tail when perched.
Voice: Main call is 'wee-tuc-tuc-tuc': song is a metallic warble.
Feeding habits: Largely insectivorous, feeding like Black Redstart.
Nest: Like that of Black Redstart, built in holes in trees or walls or on ledges in buildings.
Habitat and range: Woods, parks, orchards, mature gardens, and scrubby areas, including old quarries. A summer visitor to most parts of Europe.

1

2

3

1

Robin

Erithacus rubecula

Description: 14cm. The brown back and wings and the red face and breast distinguish it from all other species. Young have speckly brown breasts.
Voice: The commonest call is a loud and rapid 'tic-tic-tic. The song, heard throughout the year apart from a few weeks in mid-summer, is a melodious medley of high-pitched trills and warbles.
Feeding habits: Insects, spiders, and other small invertebrates are the main foods. Small seeds and soft fruits are also eaten and all kinds of foods are taken from the bird table in winter. Robins have a weakness for meal-worms and quickly learn to take them from the hand.
Nest: A cup of leaves and moss, lined with hair and fine roots. Built in dense vegetation or in a cavity: happy to use open-fronted nest-boxes.
Habitat and range: Woods, hedgerows, parks, and gardens. Common almost everywhere: few British gardens, even in towns, are without a robin. Strongly territorial for most of the year. Throughout Europe except the far north, but only a summer visitor to Scandinavia.
Similar species: Red-breasted Flycatcher has no red on its face.

2

Song Thrush

Turdus philomelos

Description: 24cm. The breast has a clear buff or sandy tinge, fading to white at the rear. Sandy underside of wing clearly visible in flight.
Voice: A beautiful flute-like song, with each note or phrase repeated several times. The commonest call is 'sip-sip-sip'.
Feeding habits: Omnivorous, taking a wide range of invertebrates and fruit. Snails are the favourite food. The shells are hammered against stones until they break and the thrush can extract the flesh. Visits bird tables in winter, but prefers feeding on ground.
Nest: A cup of grass and other plant material, lined with mud, dung, or rotten wood. Built in dense cover: often in honeysuckle or ivy in gardens.
Habitat and range: Woods, orchards, parks, and gardens: common in towns. All year in western Europe and parts of south: summer visitor elsewhere.
Similar species: Mistle Thrush is greyer and lacks sandy breast.

3

Mistle Thrush

Turdus viscivorus

Description: 28cm. Greyish brown back and wings. Underparts creamy-white with no sandy tinge on breast. The spots are less triangular and often bigger than those of Song Thrush. A white flash is visible under the wing in flight, and outer tail feathers are tipped with white.
Voice: A loud, ringing song; less varied than Song Thrush without repetition.
Feeding habits: Feeds mainly on fruit, especially yew, rowan, and ivy. Takes dried fruit from bird table, but prefers feeding on ground.
Nest: A cup of plant material, mixed with soil and lined with fine grass. Built by female, usually in fork of a tree or shrub.
Habitat and range: Woods, parks, orchards, and mature gardens: common in suburban areas. Resident in most areas as far as southern Sweden: a summer visitor further north.

1/2

3

1 Fieldfare

Turdus pilaris

Description: 25cm. Easily identified by blue-grey head, nape, and rump, contrasting with rich brown back and wings. Dark streaks on sandy breast, becoming arrow-shaped on flanks.
Voice: Main call is a raucous 'chack-chack-chack'. Song is a mixture of chuckles and squeaks, with little real structure.
Feeding habits: Omnivorous, taking insects and other invertebrates and a wide range of fruit. Visits bird tables in harsh weather.
Nest: A bulky cup of grass and other vegetation, lined with mud and then with an inner lining of fine grass. Built by female in small trees or on the ground. Does not nest in the British Isles.
Habitat and range: Open woods, parks, and gardens: farmland and other open country in winter. Resident in much of northern and central Europe, but a winter visitor to the British Isles and southern Europe.

2 Redwing

Turdus iliacus

Description: 21cm. Dark brown back and wings: a pale 'eyebrow' and a red patch on the flank: a red patch is visible under the wing in flight.
Voice: Calls include a high-pitched 'seep-seep-seep' and a harsher 'tchuk-tchuk-tchuk'. The song consists of flute-like notes, with various phrases repeated several times.
Feeding habits: Omnivorous, taking a wide range of invertebrates and fruit. Scours hedgerows for fruit in winter and also visits bird tables.
Nest: A cup of twigs and other vegetation, usually plastered internally with mud and lined with fine grass: in trees or shrubs or on the ground.
Habitat and range: Tundra, open woods, town parks, and gardens in summer: farmland, hedgerows, parks, and gardens in winter. Breeds in northern Europe, including northern Scotland, but moves south for the winter.

3 Blackbird

Turdus merula

Description: 25cm. Male jet black with a bright golden bill. Female dark brown with indistinct spotting on breast and a brown bill.
Voice: The commonest calls are 'pink-pink-pink' and a loud 'chook-chook-chook', the latter uttered when the bird is alarmed. The song is rich and flute-like, with a variety of phrases.
Feeding habits: Omnivorous, taking insects, worms, and a wide range of fruits. A regular visitor to the bird table.
Nest: A cup made with grass and other plant material, lined with mud and then further lined with fine grass or other leaves. Built by female in dense vegetation: often in honeysuckle or ivy on walls.
Habitat and range: Almost anywhere with trees or bushes: one of the commonest birds in the British Isles and present in almost every garden, even in towns. Most of Europe, but only a summer visitor in far north.
Similar species: Young starlings are sometimes mistaken for female blackbirds, but have longer bills and much shorter tails.

1

2

3

1 Starling

Sturnus vulgaris

Description: 22cm. Essentially black with green and purplish iridescence and pale spots, the latter being most obvious in winter. Bill yellow in summer, brown in winter. Wings very triangular in flight. Starlings strut over the ground and do not hop as most other garden birds do. Young birds are dull brown and sometimes confused with female blackbirds, but blackbirds have shorter bills and longer tails.
Voice: A wide range of clicking and whistling calls and songs, often mimicking other birds. Usually flaps wings while calling or singing.
Feeding habits: Omnivorous, taking insects and many other invertebrates as well as fruit: can be a serious pest in cherry orchards. A regular at the bird table in winter, often chasing smaller birds away.
Nest: An untidy cup of stalks and leaves, lined with moss and feathers. Built in holes in trees and buildings or in dense creepers.
Habitat and range: Woods, parks, and gardens. Huge numbers gather together in noisy roosts in winter, often on town buildings. Breeds nearly all over Europe except Iberia, but only a summer visitor to the north.

2 Jackdaw

Corvus monedula

Description: 33cm. Black with a conspicuous slate-grey nape.
Voice: A rather high-pitched 'chack-chack-chack', almost like a yelp, and a ringing 'keeow-keeow'.
Feeding habits: Omnivorous, taking worms, insects and many other invertebrates, nestling birds and young rodents, and plenty of fruit. Often seen digging into park lawns and roadside verges. Commonly associates with sheep to catch insects attracted by the droppings.
Nest: An accumulation of twigs, lined with wool and plant fibres. Usually built in a hole in a tree or building: often in small colonies.
Habitat and range: Open woodland, farmland, parks, churchyards, and large gardens. Most of Europe but absent from the far north.

3 Magpie

Pica pica

Description: 45cm. Instantly recognised by its black and white pattern and long tail.
Voice: Normally a harsh rattle, but emits a musical 'chook-chook-chook' in spring.
Feeding habits: Omnivorous, but with a bias towards animal food, including nestling birds and other small vertebrates as well as a wide range of invertebrates. Commonly feeds on roadside carrion.
Nest: A bulky cup of sticks and mud, lined with fine roots and other fibres and roofed with twigs. Usually built in a tree or tall hedge.
Habitat and range: Woodland margins, hedgerows, parks, churchyards, and anywhere else with scattered trees and bushes. A common town bird in the north. Resident throughout Europe.

1

2

3

1 Dunnock

Prunella modularis

Description: 14cm. Dull brown above, with greyish head and breast and streaked brown flanks. Thin bill. Often called hedge sparrow, although not a sparrow at all.

Voice: A thin, high-pitched 'tseep-tseep-tseep'.

Feeding habits: Food, consisting mainly of insects, spiders, and small seeds, is usually gathered on the ground. The bird will come to the bird table, but prefers to feed on scraps falling to the ground below.

Nest: A cup of twigs, leaves, and other vegetation, lined with hair and moss. Built by both sexes in a dense bush or hedge.

Habitat and range: Almost anywhere with plenty of cover, including heaths, parks, and gardens. One of the commonest suburban birds, taking advantage of the parks and the miles of privet hedges.

Similar species: Female House Sparrow is superficially similar but has a much stouter bill and lacks the grey breast.

2 House Sparrow

Passer domesticus

Description: 15cm. Streaky brown back and wings and a stout bill in both sexes: male has a grey crown and black bib, but female has dull brown crown and no bib.

Voice: A range of twittering chirps and cheeps, including a repetitive 'didit-didit-didit'.

Feeding habits: Takes mainly seeds, insects, and spiders, but will eat almost anything on the bird table in winter. Gregarious at all times.

Nest: An untidy mass of grass and straw, often with paper and string, lined with hair and feathers: usually wedged into a hole or on to a ledge on a building, often among ivy and other creepers. Sometimes built in a hedge, and then usually a neater, domed structure.

Habitat and range: Parks, gardens, and neighbouring fields and hedgerows: rarely far from human habitation. Throughout Europe.

Similar species: Dunnock has thinner beak. Tree Sparrow has chestnut crown.

3 Tree Sparrow

Passer montanus

Description: 14cm. Streaked brown back and wings, chestnut crown, and black bib in both sexes: a black cheek patch below and behind the eye.

Voice: A variety of chirps and cheeps, similar to those of House Sparrow but with a higher pitch: a characteristic 'tek-tek-tek' in flight.

Feeding habits: Takes insects and spiders, seeds and small fruits, and buds. Comes to bird tables, but less readily than House Sparrow.

Nest: An untidy mass of twigs and straw, lined with feathers. Usually built in holes in trees or buildings: sometimes wedged among creepers.

Habitat and range: Open woodland, orchards, parks, and large gardens, but much less tied to human settlements than House Sparrow.

Similar species: House Sparrow (above) never has black cheek patch.

1

2

3

1 Chaffinch

Fringilla coelebs

Description: 15cm. The male is easily recognised by his slate-blue head and nape, pink face and underparts, and rich brown back. The female is dull olive brown above and greyish brown below. Both sexes have a broad white shoulder flash and a narrower white wing bar.

Voice: A wide range of calls, the commonest being 'pink-pink-pink'. The song varies from place to place but is usually a cheerful rattle of about a dozen notes, often ending in what sounds something like 'cheerio'.

Feeding habits: Eats insects, spiders, buds, and seeds. A regular visitor to the bird table, although it prefers to forage on the ground below.

Nest: A neat cup of mosses, lichens, and other plant material, often bound together with spider silk, and lined with hair and feathers. Usually wedged into the fork of a tree or shrub.

Habitat and range: Almost anywhere with tree, shrubs, or hedges: often in more open country in winter. Few gardens are without it. Breeds throughout Europe, but only as a summer visitor in the far north.

2 Bullfinch

Pyrrhula pyrrhula

Description: Male is readily identified by black cap, rosy underparts, and broad white wing bar. The white rump is conspicuous in flight. Female, on the left, has a similar pattern but is much duller.

Voice: Rather quiet, the main call being a soft, whistling 'pew-pew-pew'.

Feeding habits: Primarily a seed-eater, but also eats soft fruit. Can be a serious pest in spring by eating buds of fruit trees and bushes.

Nest: A cup of twigs, moss, and lichen, lined with roots and hair. Built by female in a dense bush or hedge.

Habitat and range: Woods, hedgerows, parks, and gardens with plenty of cover. Throughout Europe except the far north and southern Iberia.

3 Goldfinch

Carduelis carduelis

Description: 12cm. The red face and golden yellow wing bars readily identify this bird.

Voice: A lilting, tinkling 'tsweet-tsweet-tsweet' and a harsher call of 'geeze-geeze-geeze'.

Feeding habits: Small seeds are the main foods, often taken from the flowerheads of dandelions and other species before they are ripe. It will take small or crushed seeds from the bird table. Also eats insects and spiders.

Nest: A neat cup of moss, grass, and other plant material, lined with wool and plant hairs. Built by female, usually in a tree or tall shrub.

Habitat and range: Woodland edges, hedgerows, orchards, parks, and gardens: often in more open country in winter. Resident throughout southern and central Europe: summer visitor to southern Scandinavia.

1

2

3

1 Greenfinch

Carduelis chloris

Description: 15cm. Olive green on the head and back, with bright yellow wing patches and a yellow rump. Female is duller and somewhat browner on the back than the male, but otherwise similar.
Voice: Most distinctive flight call is a repetitive 'chi-chi-chi'. Song is a twittering mixture of these calls and a prolonged 'dwee-ee-ee-ee'.
Feeding habits: Essentially a seed-eater, but also takes small fruits and buds: very fond of peanuts and a regular visitor to the bird table,
Nest: A bulky cup of grass and moss, lined with hair and other fibres. Usually built in the fork of a tall bush or tree, especially a conifer.
Habitat and range: Woods, hedgerows, orchards, parks, and gardens: often moves to more open country in winter. All Europe except the far north.

2 Siskin

Carduelis spinus

Description: 12cm. Male has a black cap, small black bib, clear yellow breast, a yellow rump, and streaked flanks. Female lacks the black cap and bib and is largely greyish yellow with a heavily-streaked breast. Both sexes have yellow flashes on the sides of the tail.
Voice: Commonest call is 'tsew-tsew-tsew'. Song is a musical twitter, often delivered in flight.
Feeding habits: Primarily a seed-eater, with a particular liking for conifers and alders: readily takes peanuts from bird-feeders and also eats insects and spiders.
Nest: A cup of small twigs, moss, and grass, lined with feathers and various fibres. Usually built high up in a conifer.
Habitat and range: Breeds mainly in wooded areas, but increasingly common in parks and gardens in winter. Resident over much of central Europe, including the British Isles: a summer visitor in the far north and a winter visitor in much of the south.
Similar species: Serin has no black cap and no yellow on sides of tail.

3 Serin

Serinus serinus

Description: 12cm. Strongly streaked, yellowish green back and wings, with a bright yellow rump. Male has bright yellow face and breast and streaked flanks. Female is much duller, with a streaked breast.
Voice: Twitters musically in flight: song is a cheerful cascade of whistling notes.
Feeding habits: Essentially a seed-eater, but also takes small insects and spiders. Feeds mainly on ground.
Nest: A neat cup of moss and other plant materials, lined with hair and feathers. Built in trees and bushes.
Habitat and range: Woodland edges, olive groves, and vineyards: common in towns and gardens in southern Europe, where it is resident. A summer visitor elsewhere. A scarce visitor to Britain until recently, but now breeding in small numbers. Absent from most of Scandinavia.
Similar species: Siskin (above) has bright yellow flashes on tail.

2

3

1 Common Wall Lizard

Podarcis muralis

Description: Head and body (as far as hind legs) up to 7.5cm: tail up to about 15cm. A rather variable lizard, but generally brown or grey with vertical black and white bars on the tail. Males usually have dark spots on the back and both sexes are sometimes tinged with green.
Food and habits: Eats flies and other insects and spiders. A diurnal, climbing species commonly basking on the tops of walls and even on tiled roofs. Dormant in winter in cooler parts of its range.
Habitat and range: Rocks, walls, and terraces, usually in rather dry, open places. By far the most frequent lizard around houses and common in many towns and villages. Widespread in southern and central Europe, but absent from British Isles (except Channel Islands) and most of Iberia.
Similar species: Several similar lizards live in southern Europe, especially in the Balkans, but are less likely to be seen in gardens.

2 Slow-worm

Anguis fragilis

Description: Up to 50cm long, this legless lizard is usually brown or grey, often with a coppery tinge, and very smooth and shiny. Female may have a dark line along the back: male may have a few blue spots. Easily distinguished from a snake by its more oval eye and movable eyelids.
Food and habits: Feeds largely on slugs, but also eats worms and insects. Sometimes basks in sunshine, especially early or late in the year, but spends most of the day under logs and stones – particularly planks and other flat objects that transmit the sun's warmth to the underside. Hunts mainly in the evening and after rain. Dormant in winter.
Habitat and range: Well-vegetated and fairly damp areas, including roadside verges, orchards, grassy shrubberies, and the wilder gardens. Most of Europe except Ireland and the far north.

3 Smooth Newt

Triturus vulgaris

Description: Up to 11cm long, with smooth skin and a tail accounting for about half the length. The female is dull yellowish brown above and the underside is cream with a bright orange band down the centre and numerous dark spots on throat and belly. Outside the breeding season the male is like the female, but in the breeding season he becomes brightly coloured, as illustrated, with a wavy crest on both tail and body.
Food and habits: Eats a wide range of invertebrates on land and in the water. Also eats frog spawn and small tadpoles in spring. Eggs are laid singly on water plants in spring, after an elaborate courtship easily watched in garden ponds. The adults usually spend the rest of the year in damp places on land. They become dormant in winter.
Habitat and range: Lives in a wide variety of damp situations on land: common in many mature gardens, especially if there is a pond. Most of Europe except the far north and the south-west.
Similar species: Palmate Newt, rare in gardens, lacks throat spots.

1

2

3

1 Common Frog

Rana temporaria

Description: Up to 10cm long. Ground colour of upperside ranges from greyish brown, through olive-green, to yellow or brick-red. Underside ranges from white to yellow, speckled with brown. The dark patches on the back vary, but there is always a dark patch, enclosing the eardrum, just behind the eye. The legs usually have dark transverse bands. There is a prominent skin fold on each side of the body.
Food and habits: Eats slugs, worms, snails, and many other invertebrates on land, but does not feed in the water. Spends most of its time in damp places on land and moves to ponds and ditches to breed in spring. Frogs are very noisy at this time, when the croaking males may sound like the roar of distant motor cycles. The gelatinous egg masses (frogspawn) give rise to tadpoles that turn into tiny frogs in about three months. Winter is passed in a dormant state, often in mud at the bottom of ponds.
Habitat and range: Breeds in ponds and other still or slow-moving waters: garden ponds are important habitats now that many farm and village ponds have disappeared. All Europe except the far south.
Similar species: None in the British Isles. Agile Frog and Moor Frog on the continent often have yellow flanks, but rarely occur in gardens.

2 Common Toad

Bufo bufo

Description: Up to 15cm, but usually much less. Greyish brown with little or no pattern: sometimes yellowish or brick coloured. The warty skin distinguishes it from the Common Frog. The pupil is horizontal.
Food and habits: Eats the same food and behaves in much the same way as the Common Frog, although it tends to walk rather than hop. The eggs are laid in long strings attached to water plants.
Habitat and range: Breeds in still and slow-moving waters, but occupies all kinds of habitats outside the breeding season, often far from the breeding sites. Most of Europe except Ireland and the far north.
Similar species: Midwife Toad is smaller and has vertical pupils.

3 Midwife Toad

Alytes obstetricans

Description: Head and body usually under 5cm: grey, olive green, or brown, with conspicuous warts and sometimes with dark spots. The eyes are prominent, with vertical pupils.
Food and habits: Eats slugs and snails, insects, and other invertebrates. Adult is essentially terrestrial and nocturnal. Emits a shrill, whistling note every two or three seconds – 'poo-poo-poo'. Pairing takes place on land and the female lays a string of large creamy eggs that are wrapped around the legs of the male. He dips them in the water from time to time and enters the water when it is time for the eggs to hatch.
Habitat and range: A wide range of habitats, not necessarily near water. Often hides by day under logs and in holes in old walls. France, Iberia, and western parts of Switzerland and Germany.
Similar species: Common Toad is bigger, with horizontal pupils.

1

2

3

1 *Gryllomorpha dalmatina*

Description: Up to 18mm, with an ovipositor up to 15mm in the female. Both sexes have two hairy 'tails'. This wingless cricket is usually pale brown with darker spots, although sometimes entirely dark except for the pale cross on the thorax.
Food and habits: An omnivorous insect, consuming vegetable debris and a variety of small animals. Hides under stones or in crevices by day and emerges to feed at nightfall. Adults are found mainly in the autumn.
Habitat and range: Damp places, including cellars, old walls, and stone out-buildings. Southern Europe.

2 House Cricket

Acheta domesticus

Description: Up to 20mm, with an ovipositor up to 13mm in the female. Greyish or yellowish brown, with darker markings at the front. The hindwings are tightly rolled when not in use and project from the rear like an extra pair of tails between the bristly cerci. The female, with her needle-like ovipositor, appears to have five tails.
Food and habits: An omnivorous, nocturnal scavenger. Flies well. The male emits a shrill, whistle-like song. Adults can be found throughout the year.
Habitat and range: Found mainly in heated buildings and rubbish dumps although it commonly moves into gardens and the surrounding countryside in hot summers. A native of North Africa, it is now found in and around human habitation nearly all over the world.

3 Dark Bush-cricket

Pholidoptera griseoaptera

Description: 12–20mm. The upper surface ranges from pale brown, through chestnut, to almost black, although the top of the head and thorax are always light brown. The underside is greenish yellow. The male, illustrated below, has short, oval wings perched on top of the body, but cannot fly. The female, pictured here, is virtually wingless. She has a curved, sabre-like ovipositor up to 10mm long.
Food and habits: Omnivorous, eating a variety of small insects and plant matter. Active mainly in the evening, when the male advertises his presence with short, high-pitched chirps – 'psst-psst' – but can be seen and heard at most times of the day. Adult June–November.
Habitat and range: Rough vegetation of woodland clearings, roadside verges, hedgerows, and garden shrubberies: particularly fond of brambles and nettle beds. Most of Europe, but rare in Scotland and Ireland.

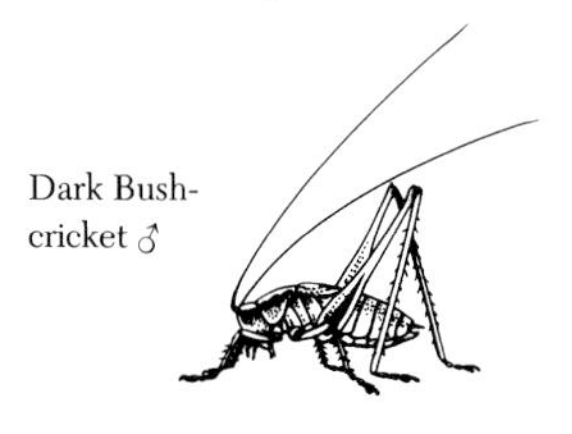
Dark Bush-cricket ♂

1

2

3

1 Great Green Bush-cricket

Tettigonia viridissima

Description: Up to 60mm from the head to the tip of the folded wings. The female's slightly down-curved ovipositor is about 20mm and just reaches the wing-tips. Bright green, with a brown streak along the top.

Food and habits: Largely carnivorous, eating many other insects, including grasshoppers. May bite if handled. Active mainly from late afternoon until after midnight. The male produces a loud, strident song somewhat like the sound of a sewing machine. Adult July–October.

Habitat and range: Rough vegetation, including hedgerows, shrubberies, and large herbaceous beds. Most of Europe, but absent from Scotland, Ireland, and the far north.

2 Oak Bush-cricket

Meconema thalassinum

Description: Up to 15mm from the head to the tip of the folded wings. The female's curved ovipositor is about 9mm and extends well beyond the wings. The male has a pair of slender claspers at the rear. Pale green, often with a bluish tinge, and with a yellow or brownish streak on the top.

Food and habits: Feeds mainly on aphids and other small insects, but may also nibble leaves. Strictly nocturnal. Both sexes fly well and often come to lighted windows. The male has no song and attracts a mate by drumming on leaves with his back feet. Adult July–November.

Habitat and range: Lives mainly in trees, including garden apples and plums. Most of Europe, but absent from most of Scandinavia.

3 Speckled Bush-cricket

Leptophyes punctatissima

Description: 10–15mm. The male, pictured here, is bright green, densely speckled with black and with a narrow brown stripe along the back. The female, illustrated below, has a broad, strongly-curved ovipositor about 7mm. Both sexes are flightless, with very short brown or green wings perched on top of the body.

Food and habits: Largely vegetarian, although it does eat some aphids. Active day and night, but very well camouflaged and not easy to see. The male's song is a barely audible scratching sound. Adult July–November.

Habitat and range: Dense vegetation, including roadside verges, hedgerows, shrubberies, and herbaceous beds. Most of Europe except the far north, but rare in the northern half of the British Isles.

Speckled Bush Cricket ♀

1

2

3

1 Common Earwig

Forficula auricularia

Description: 10–15mm, excluding the pincers. The latter are 4–9mm in the male, strongly curved and with a flat base. Female pincers (pictured here) are 4–5mm and almost straight. The body is shiny brown, with the hindwings projecting as pale triangles from under the short forewings.
Food and habits: Mainly vegetarian, eating a wide range of living and dead plant material, but also taking some insect food. Sometimes damages flowers. Largely nocturnal, hiding under loose bark and in other crevices by day. Both sexes can fly, but rarely do so. Female looks after her eggs and young and family groups are often disturbed under flower-pots and other objects in the garden. Young earwigs always have straight, slender pincers. The white or cream earwigs that are often unearthed in the garden are not different species – simply individuals that have just changed their skins. Adult all year, but usually dormant in winter.
Habitat and range: Abundant almost everywhere, in the house as well as in the garden. Often in trees, commonly nestling in clusters of apples in late summer. Throughout Europe. The only earwig commonly seen in the British Isles.
Similar species: Several other earwigs are superficially similar, but few have projecting hindwings.

2 Small Earwig

Labia minor

Description: 5mm, with pincers up to 2.5mm. The smallest European earwig, it is dull brown and rather hairy, with a blackish head. The hindwings project as small triangles from under the very short forewings.
Food and habits: Omnivorous, feeding on decaying plant and animal matter and also taking some animal food. Both sexes fly well, usually in the evening. Breeds in manure and compost heaps, where active adults can be found throughout the year. Exhibits parental care as Common Earwig.
Habitat and range: Abundant wherever there is decaying matter, including manure heaps, rubbish dumps, and garden compost heaps. All Europe.

3 Hop-garden Earwig

Apterigida media

Description: 6–12mm: pincers 2–5mm, straight in the female but gently curved and with a prominent tooth in the male. Pale, clear brown, with very short foreings and no hindwings.
Food and habits: Largely vegetarian, sometimes damaging flowers. Mainly nocturnal, often hiding in flowers and seed capsules by day. Adult all year, but usually dormant in winter. Probably exhibits parental care like Common Earwig.
Habitat and range: Shrubs and tall herbs, including garden borders. Once common in Kentish hop fields. Most of Europe except the far north, but in the British Isles it is confined to south-east England and East Anglia.

1

2

3

1 Pied Shield Bug

Sehirus bicolor

Description: 5–7mm, with spiny legs. Often mistaken for a beetle – especially a 'black-and-white ladybird' – but the membranous tips of the forewings clearly show that it is a bug and not a beetle. Like all bugs, it has a piercing beak for sucking up sap and other liquids.

Food and habits: Feeds on deadnettles and related plants, especially on the developing fruits. Adult most of the year, but hibernates in the soil in winter.

Habitat and range: Most well-vegetated habitats, including the bases of garden walls and hedges. Most of Europe, but not Scotland or Ireland.

2 Hawthorn Shield Bug

Acanthosoma haemorrhoidale

Description: 15mm. Thorax is distinctly triangular, with a red band at the rear. The shield-shaped scutellum is also bounded by red.

Food and habits: Feeds on the buds, leaves, and fruits of hawthorn and various other deciduous trees. Can often be seen basking on walls in autumn before going into hibernation in moss or leaf litter.

Habitat and range: Woodland margins and hedgerows, including garden hedges and shrubberies. Much of Europe, but absent from Scotland.

3 Common Green Shield Bug

Palomena prasina

Description: 10–15mm. Bright green in spring and summer, but becomes bronzy green just before hibernation in autumn. The sides of the thorax are very slightly concave and the membrane at the wing-tip is quite dark. Nymph is pale green, very flat, and almost circular.

Food and habits: Feeds on a wide range of trees, shrubs, and tall herbaceous plants, sometimes damaging beans. Adult most of the year, but hibernates in leaf litter in winter.

Habitat and range: Woodland edges and clearings, hedgerows, shrubberies, and herbaceous borders. Most of Europe.

Similar species: P. viridissima, absent from British Isles, has slightly convex edges to the thorax. *Nezara viridula* has a paler membrane.

4 *Nezara viridula*

Description: The nymph, pictured here, is up to 10mm and quite unmistakable. The adult, 10–15mm, is bright green and very like Common Green Shield Bug, but with a pale brown membrane at the wing-tip.

Food and habits: Feeds on a wide range of herbaceous plants and causes considerable damage to peas and potatoes. Also on various trees and shrubs. Winter is passed as an adult or fully-grown nymph.

Habitat and range: Occurs in almost any well-vegetated wild or cultivated habitat. Throughout southern Europe, occasionally carried to the British Isles and other areas in produce.

Similar species: Common Green Shield Bug nymph is plain green.

1/2

3

4

1 Forest Bug

Pentatoma rufipes

Description: 12–15mm: dark brown with reddish legs. Thorax almost rectangular with a backward-pointing spine on each 'shoulder'.
Food and habits: An omnivorous insect, taking sap from buds, leaves, and fruits and also attacking other small insects. Adult June–October.
Habitat and range: Occurs on a wide range of deciduous trees and shrubs and often common in orchards and garden shrubberies. Most of Europe.
Similar species: Several other bugs have a similar coloration, but none has the rectangular thorax and 'square-shouldered' appearance.

2 *Zicrona caerulea*

Description: 6–8mm. Dark green, blue, or violet.
Food and habits: A carnivorous bug, attacking the eggs, larvae, and adults of other insects and sucking the juices from them. Many garden pests, including the grubs of the Colorado Beetle and the caterpillars of the cabbage white butterflies, are among its victims. It also destroys numerous vine pests on the continent. Adult throughout the year, although it hides away in the coldest months.
Habitat and range: A wide range of wild and cultivated habitats, wherever it can find insect victims, although it is not common on cultivated land in the British Isles. Most of Europe except Ireland.

3 *Eurydema dominulus*

Description: 6–10mm. Usually red and black, but the red sometimes replaced by yellow or cream and the dark spots may be metallic green.
Food and habits: Attacks various members of the cabbage family and is a pest of most cultivated brassicas, including turnips. Adult for most of the year, but hides away in leaf litter and other debris in winter.
Habitat and range: Most common on cultivated land, but also on wild crucifers in hedgerows and waste land. Much of Europe, but a rare insect in the British Isles.

4 Brassica Bug

Eurydema oleracea

Description: 6mm, with a very variable appearance. The ground colour is usually metallic green or blue and the paler spots range from cream, through yellow, to red.
Food and habits: Feeds mainly on members of the cabbage family and is a serious pest of most cultivated brassicas. It also attacks vines and is sometimes found on potatoes. Adult most of the year, but hides in debris for the winter.
Habitat and range: Mainly on cultivated land, but also in hedgerows and waste places. Most of Europe, but uncommon in the British Isles.

1

2

3/4

1 Fly Bug

Reduvius personatus

Description: About 17mm: chocolate brown to black and rather bristly.
Food and habits: Feeds on many other small insects, including bed bugs in the house, but will stab people with its sharp beak if handled. Also squeaks when handled by rubbing tip of beak against thorax. Nocturnal: often attracted to lighted windows. May–September.
Habitat and range: Wooded and rocky places with plenty of crevices, but most common in and around human habitation: hides in sheds and other buildings by day. Much of Europe but uncommon in the British Isles (south only).

2 Common Flower Bug

Anthocoris nemorum

Description: 3–4mm: very shiny, with a conspicuous black spot near the middle of each greyish brown forewing. There are no veins in the membranous area at the tip of the wing.
Food and habits: A useful garden inhabitant, feeding on aphids, red spider mites, and other small creatures, although it will also pierce fingers and draw blood if handled. Adult all year, but dormant in winter – often under loose bark or in clumps of grass.
Habitat and range: On trees, shrubs, and herbaceous plants in almost any habitat – mainly on the leaves, despite its name. Most of Europe.
Similar species: Several bugs are superficially similar, but they are not shiny and they usually have veins in the wing membrane.

3 Tarnished Plant Bug

Lygus rugulipennis

Description: 4–6mm: dull yellowish brown and rather hairy, with a rust-coloured tinge. There are two pale yellow triangles towards the rear.
Food and habits: A common garden pest, attacking potatoes and many other crops and flowers: also common on stinging nettle. Causes white spotting on leaves and also attacks fruits. Adult all year, but spends the winter in leaf litter and other debris: most common in late summer.
Habitat and range: Well-vegetated habitats of all kinds. Most of Europe.

4 Common Green Capsid

Lygocoris pabulinus

Description: 5–7mm: plain green with a greyish membrane at the wing-tip and slender pale brown spines on the legs. A very narrow collar just behind the head is visible with a lens.
Food and habits: Feeds on trees and shrubs in early spring and herbaceous plants in the summer. Potatoes and soft fruit, especially raspberries and gooseberries, are among the many plants attacked: affected fruits bear blemishes or even fail to grow properly on the affected side. May–October. Overwinters in the egg stage on trees and shrubs.
Habitat and range: Well-vegetated habitats of all kinds. Most of Europe.
Similar species: Potato Capsid has two black spots just behind the head.

1

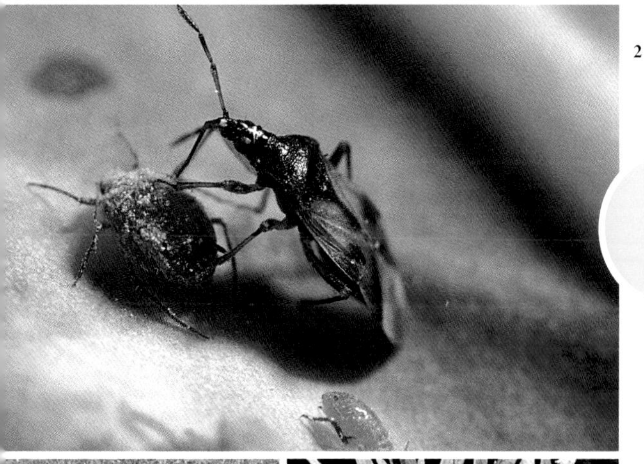

2

3/4

1 Buffalo Treehopper

Stictocephalus bisonia

Description: About 10mm: bright green with a sharp horn on each side of the body. The thoracic shield (pronotum) ends in a sharp spine.
Food and habits: Feeds on the sap of a wide range of shrubs and herbaceous plants: mostly on young shoots. If disturbed it often scuttles round to the other side of the stem, but it can also jump and fly well. It can damage apples and other fruit trees by cutting slits in the young shoots and laying its eggs there. Adult July–September.
Habitat and range: Waste ground, hedgerows, shrubberies, and herbaceous borders. An American insect now well established in Southern and Central Europe, mainly in the west. Absent from British Isles.

2 Common Froghopper

Philaenus spumarius

Description: 6mm: some shade of brown with a very variable pattern of light or dark streaks or spots. The head is sometimes white. Also called a spittlebug, it is one of several similar jumping bugs responsible for the 'cuckoo-spit' that appears on plants in spring.
Food and habits: Takes sap from many woody and herbaceous plants. The young bugs surround themselves with white froth, produced by pumping air into a fluid exuded from the rear end. The froth prevents the delicate green nymphs from drying out and also protects them from some of their enemies. Adult June–September, passing the winter in the egg stage.
Habitat and range: Any well-vegetated habitat. All Europe except the far north.

3 Potato Leafhopper

Eupteryx aurata

Description: about 4mm, with a variable black and yellow pattern, the yellow often being tinged with orange.
Food and habits: Attacks a wide range of herbaceous plants, including stinging nettles and mint as well as potatoes. Its saliva destroys chlorophyll and causes pale patches to develop around its feeding punctures. Infestations are rarely serious in the garden, although large areas of leaf may be destroyed if the bug is present in large numbers. Adult May–November.
Habitat and range: Waste ground, hedgerows, gardens and most other well-vegetated habitats. All Europe except the far north.

4 Rhododendron Leafhopper

Graphocephala fennahi

Description: 8–10mm, readily identified by its red-streaked wings.
Food and habits: Takes sap from rhododendron bushes. Also damages flowers by laying eggs in the buds and by carrying the fungus causing bud-blast disease. Adult June–October, passing the winter in the egg stage.
Habitat and range: An American insect now well-established on wild and cultivated rhododendrons in southern England.

1

2/3

4

Aphids

Aphids are small, sap-sucking bugs, with or without wings. They are familiar to gardeners as greenfly and blackfly, although there are dozens of species in our gardens. Each has its preferred host plant and many have complicated life cycles involving two different plants at different times of year. Winged forms tend to appear as the populations increase, and they fly off to start colonies on fresh plants. Reproduction during the summer is mainly by virgin birth or parthenogenesis. Males normally appear only in the autumn, and this is when mating and egg-laying normally occur. The sugary honeydew that aphids exude is eagerly sought by many other insects. A black fungus also grows on it and disfigures the plants. Production of honeydew is the aphids' way of expelling excess sugars taken in with the sap.

1 Woolly Aphid

Eriosoma lanigerum

Description: 1–2mm: purplish brown, with or without wings. The body is usually concealed by fluffy white strands of wax.

Food and habits: Feeds on apples and related trees, forming dense, woolly clusters on young twigs or on the bark of older branches, especially where the trees have been wounded. Young twigs become swollen and cracked. Active colonies are most obvious in summer. Reproduction is almost entirely by parthenogenesis; eggs are rare. Only young aphids survive the winter, either hiding in bark crevices or, more rarely, feeding on the roots.

Habitat and range: Mainly orchards and gardens. An American insect, now widely established in Europe: often called American Blight.

2 Cabbage Aphid

Brevicoryne brassicae

Description: Up to 2mm. Winged forms are black and dark green: wingless forms are green with a thick mealy coating.

Food and habits: Forms dense masses on brassica leaves in spring and summer and causes severe damage, especially when it attacks young plants. Whole plants may wither and die. Winter is usually passed as eggs but adults can survive mild winters. Old brassica stumps left in the garden are the main sources of reinfestation in the spring.

Habitat and range: Waste ground and cultivated land – wherever plants of the cabbage family can be found. Throughout Europe.

3 Black Bean Aphid

Aphis fabae

Description: About 2mm: black or olive green, with or without wings.

Food and habits: In summer forms dense clusters on young shoots of beans, docks, spinach, beet, nasturtiums and others, distorting leaves and weakening plants by removing sap. Carries damaging viruses. Winter is passed in the egg stage on spindle, *Viburnum*, and *Philadelphus* bushes.

Habitat and range: Almost anywhere with suitable host plants. All Europe.

Similar species: Several other black aphids can be found in the garden, but *A. fabae* is the only one that feeds on beans.

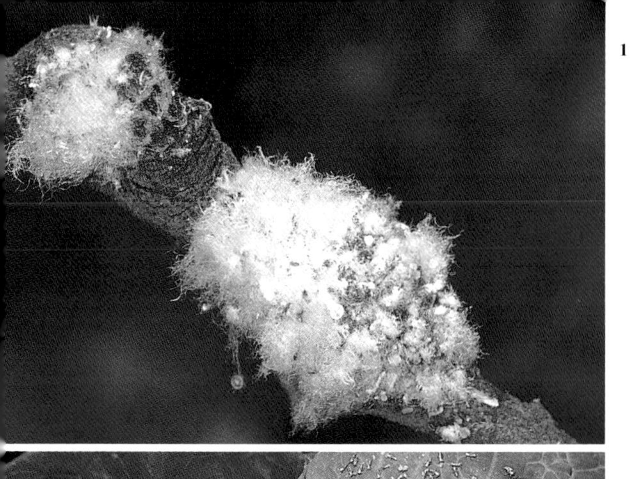

1

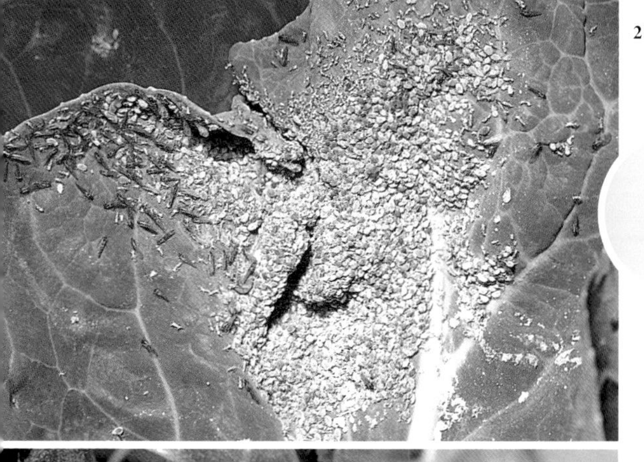

2

3

1 Cabbage Whitefly

Aleyrodes proletella

Description: 2–3mm with waxy white wings. It resembles a tiny moth, but it is actually a sap-sucking bug.
Food and habits: Lives in clusters on the undersides of brassica leaves, which it coats with honeydew. Flies off in clouds if the leaves are shaken. Heavy infestations cause serious crop damage. Active all year.
Habitat and range: Fields and gardens. All Europe.
Similar species: Greenhouse Whitefly attacks tomatoes, fuchsias, and many other greenhouse and house plants.

2 Green Lacewing

Chrysopa pallens

Description: 15–20mm: wingspan 30–40mm. The body is bright green and the eyes are brassy or golden – giving the insect its alternative name of 'goldeneye'. The veins on the delicately netted wings are also bright green. Seven tiny black spots on the head distinguish this from several similar species, but they can be seen only with a lens.
Food and habits: A carnivorous insect feeding mainly on aphids – and therefore an ally of the gardener. Often comes to lighted windows at night. May–August. The eggs, each with a slender stalk, are laid under leaves. The bristly, shuttle-shaped larvae (2a) also feed on aphids, from which they suck the juices through large, hollow jaws. Winter is passed as a fully grown larva in a silken cocoon.
Habitat and range: Woods, hedgerows, gardens, and well-vegetated places of all kinds. Most of Europe, but absent from Scotland and the far north.
Similar species: Many other green lacewings are likely to be found in the garden. Most are very similar, although some are much bluer than *C. pallens*, especially on the veins. *Chrysoperla carnea* becomes flesh-coloured in the autumn, at which time it often enters shed and houses for hibernation. It becomes green again in the spring. The larvae of many lacewing species camouflage themselves by covering their bodies with the empty skins of their aphid victims.

3 Scorpion Fly

Panorpa communis

Description: Up to 15mm: wingspan about 30mm, with narrow, spotted wings and a sturdy beak under the head. Only the male has the up-turned tail that gives the insect its name. The female abdomen is straight.
Food and habits: A weak-flying scavenger, feeding mainly on dead animal matter: also nibbles fruit and sometimes 'steals' flies from spider webs. Mainly May–August. Despite its name, the insect is quite harmless. The caterpillar-like larvae live as scavengers in the soil.
Habitat and range: Woods, hedgerows, and shady gardens: common in and around nettle beds. All Europe except the far north.
Similar species: Two very similar species occur in the British Isles, and several others on the Continent. The spot patterns vary and accurate identification depends on detailed examination of the abdomen.

1

2

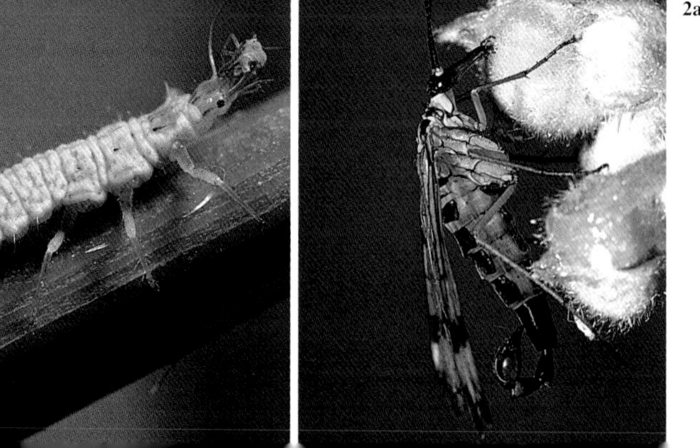

2a/3

1 Violet Ground Beetle

Carabus violaceus

Description: 20–35mm: largely black, with bright violet reflections around the thorax and wing-cases or elytra. The latter are more or less oval, with no sharp angles at the front, and have an almost smooth surface.
Food and habits: A flightless, fast-running, nocturnal predator, feeding on slugs and many other ground-living invertebrates. Hides under logs and stones by day, and also in damp sheds and other out-buildings. The larva (1a) is also predatory, although less agile than the adult.
Habitat and range: Woods, hedgerows, gardens, and waste ground, wherever there is food, shelter, and moisture. All Europe.

2 *Carabus nemoralis*

Description: 20–30mm, ranging from bronze to brassy green and often tinged with violet on the sides. The elytra bear fine ridges and rows of small pits. The female is much less shiny than the male, pictured here.
Food and habits: A flightless, fast-running, nocturnal predator with habits similar to those of the Violet Ground Beetle (above).
Habitat and range: Almost any habitat, but on the Continent it seems to be most common in parks and gardens. All Europe except the far north.
Similar species: There are several superficially similar species, but most of them are woodland insects.

3 Strawberry Beetle

Pterostichus madidus

Description: 12–20mm: shiny black with chestnut-coloured legs. The thorax is almost square, with blunt rear corners.
Food and habits: A flightless, nocturnal predator, feeding on slugs and other invertebrates, although it also attacks strawberries and other soft fruit.
Habitat and range: Farmland, parks and gardens, and most other kinds of open country. Throughout Europe.
Similar species: There are many superficially similar beetles, but most have black legs and pointed rear corners on the thorax.

4 Devil's Coach-horse

Staphylinus olens

Description: 20–30mm: velvety black with short, almost square elytra that leave most of the abdomen uncovered. These short elytra are characteristic of the group known as rove beetles.
Food and habits: A nocturnal predator, feeding on slugs and many other invertebrates. When alarmed, it opens its ferocious-looking jaws and raises its tail end – giving rise to its alternative name of cock-tail.
Habitat and range: Woods, hedgerows, parks, and gardens: often in damp sheds and other out-buildings. Most of Europe.

1/2

1a

3/4

1 Cockchafer

Melolontha melolontha

Description: 20–30mm, with a black thorax and truncated rust-coloured elytra that leave the pointed tip of the abdomen exposed. The antennae, larger in the male (pictured here) than the female, can open out like a fan.
Food and habits: Chews the leaves of various deciduous trees and shrubs, often swarming around them at dusk and commonly crashing into lighted windows. May–July: often called a may-bug. The plump, white, C-shaped larva (illustrated below) lives in the soil for 3 years and causes severe damage to cereals and other field crops.
Habitat and range: Woodland margins, hedgerows, parks, and gardens, although the females usually lay their eggs in more open habitats. Most of Europe except northern Scandinavia.
Similar species: Summer Chafer (below) has a brown thorax.

2 Summer Chafer

Amphimallon solstitialis

Description: 14–18mm: entirely brown and very hairy, with truncated elytra. The antennae have only 3 small flaps (cockchafer antennae have at least 4).
Food and habits: Adults swarm around many deciduous trees and shrubs at night and nibble the leaves. They often come to lights. The larvae feed on roots, especially grass roots, and resemble those of the Cockchafer, but they mature in only 2 years.
Habitat and range: Hedgerows, fields, parks, and gardens. Most of Europe.
Similar species: Several similar, but generally less hairy species occur on the Continent. Cockchafer (above) has a black thorax.

3 Garden Chafer

Phyllopertha horticola

Description: 7–12mm: head and thorax metallic green or black, with scattered long hairs. The truncated elytra are pale to dark brown, sometimes with a greenish iridescence.
Food and habits: Adults feed on the leaves of various deciduous trees and shrubs, often causing damage to orchard trees. They fly mainly by day. June–July. The larvae feed on grass roots, taking 2–3 years to mature.
Habitat and range: Most common in rough grassland, but also in orchards, parks, and gardens with extensive lawns. Most of Europe except the far north.

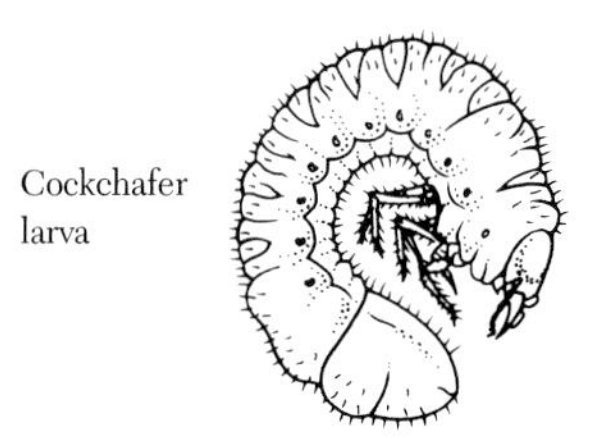

Cockchafer larva

1

2

3

1 Rose Chafer

Cetonia aurata

Description: 14–20mm: elytra rather flattened, green or bronze with a variable amount of transverse white streaking. Coppery red underneath with a rounded club-like spur between the middle pair of legs.
Food and habits: Feeds mainly on flowers, especially the stamens, and sometimes damages roses. Flies by day with a loud buzzing noise. May–August. The larvae develop in decaying wood and leaf litter. Winter is passed as a newly-formed adult, still in its pupal case.
Habitat and range: Woodland margins, hedgerows, scrub, and gardens. Southern and Central Europe: confined mainly to the south in the British Isles, where the beetle has become rare in recent years.
Similar species: C. cuprea is very similar but has a square-ended spur between its middle legs. It is a more northerly species, less often found in gardens.

2 Click Beetle

Agriotes lineatus

Description: 7–10mm and almost bullet-shaped. Dull yellowish brown, with darker stripes on the elytra: thorax sometimes dark brown.
Food and habits: Adult chews grasses and various flowers, especially the pollen-carrying stamens. Mainly May–July. It is one of a large group known as click beetles because, when they fall on their backs, they flick themselves into the air with a loud click – and carry on doing this until they land right-way-up. The larva is one of the wireworms, living in the soil for two or more years and causing severe damage to crop roots.
Habitat and range: Grassland of all kinds, arable fields, parks, and gardens. Throughout Europe except the far north.
Similar species: There are several very similar species in the British Isles and elsewhere in Europe.

3 Pollen Beetles

Meligethes species

Description: 1–3mm: mostly black with green, blue, or violet iridescence and clubbed antennae. About 60 species live in Europe, most of them very similar and very hard to distinguish.
Food and habits: Adults feed mainly on pollen, but may nibble other parts of flowers as well. Wallflowers and other spring-flowering crucifers are regularly visited, and so are newly opened poppies. The larvae chew leaves and each species has its preferred host. Adults may also nibble leaves and can severely damage seedlings. Adults are active from spring to autumn and spend the winter dormant in soil or litter.
Habitat and range: Wherever flowers grow. Throughout Europe.
Similar species: Flea beetles (p.82) have slender antennae.

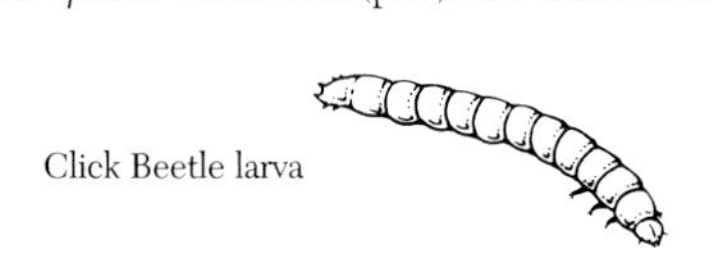

Click Beetle larva

1

2

3

Ladybirds

Ladybirds are among the most popular beetles because of their bright colours and their aphid-eating habits, although not all species are carnivorous: several eat mildews and some are actually plant pests, especially in southern Europe. Most ladybirds are strongly domed and more or less hemispherical. Their bright colours warn birds that they are distasteful: many emit a pungent, staining fluid if handled. Most ladybirds pass the winter as dormant adults, sometimes in dense masses, although some of the vegetarian species may remain active in mild winters.

1 7-Spot Ladybird

Coccinella 7-punctata

Description: 5–8mm, with 7 black spots on its bright red or orange elytra. The spot at the front is in the mid-line and shared by both elytra, and there are 3 spots further back on each side.

Food and habits: A voracious aphid-eater, scouring a wide range of plants for food. Active from early spring to autumn. The larva (1a), steely blue with yellow or cream spots, also preys on aphids. Winter is passed, singly or in small groups, in leaf litter or other sheltered places close to the ground.

Habitat and range: Abundant in most well-vegetated habitats. All Europe.

2 2-Spot Ladybird

Adalia bipunctata

Description: 3–6mm, with black legs and usually just one black spot on each elytron, although it varies a great deal. Some specimens have 4 black spots and some have black bands across the elytra. In northern areas the insect often has black elytra with red spots. The black ground colour is believed to help with heat absorption in what is usually a cooler and cloudier environment.

Food and habits: Eats aphids on woody and herbaceous plants from spring to autumn. The larva is very similar to that of the 7-spot Ladybird. The adults often enter sheds and houses to pass the winter, sometimes forming clusters of a thousand or more individuals. Overwintering sites are commonly used by successive generations year after year. Other sites include bark crevices and holes in walls, usually well above the ground.

Habitat and range: Abundant in most well-vegetated habitats. All Europe.

3 10-Spot Ladybird

Adalia 10-punctata

Description: 3–5mm, normally orange or red with yellowish brown legs and 5 black spots on each elytron, although the species is very variable. The spots sometimes link up to form a black lattice, and some individuals have just 2 red spots on a black background.

Food and habits: Feeding habits and larvae are like those of the 2-spot Ladybird, but winter is usually passed in leaf litter on the ground.

Habitat and range: Mainly woods and hedgerows: prefers trees and shrubs and is less common in gardens than the 2-spot Ladybird. All Europe.

1/2

1a

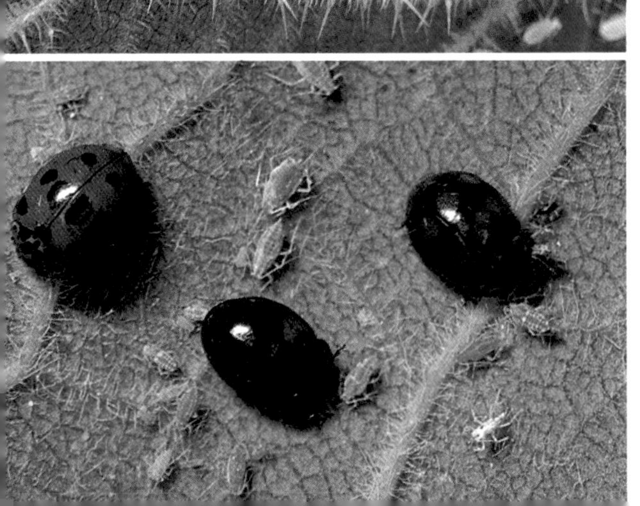

3

1 14-Spot Ladybird

Propylea 14-punctata

Description: 3–5mm, typically with 7 black spots on each cream or yellow elytron, but the spots often link up and are sometimes so extensive that the insect looks black with small yellow patches.

Food and habits: An aphid-eater abundant on a wide range of woody and herbaceous plants from spring to autumn. Winter is passed in debris or in clumps of grass or other plants close to ground level.

Habitat and range: Almost any well-vegetated habitat: especially common in hedgerows and on waste ground. Throughout Europe.

2 Cream-spot Ladybird

Calvia 14-guttata

Description: 4–6mm, with 7 pale yellow spots on each orange or rust-coloured elytron.

Food and habits: An aphid-eater active throughout the summer, mainly on trees and shrubs, including garden fruit trees. Winter is passed in leaf litter and other debris or in bark crevices.

Habitat and range: Woods, hedgerows, orchards, and garden shrubberies. Throughout Europe.

3 22-Spot Ladybird

Psyllobora 22-punctata

Description: 3–4mm, lemon yellow with 10 or 11 small black spots on each elytron.

Food and habits: A mildew eater, frequenting a wide range of shrubs and herbaceous plants, especially umbellifers: not uncommon on gooseberry bushes in gardens. Active April–August. Winter is passed in litter and bark crevices close to the ground, but the beetle may be active in mild weather.

Habitat and range: Hedgerows and other well-vegetated areas, including herbaceous borders and shrubberies. Throughout Europe.

4 24-spot Ladybird

Subcoccinella 24-punctata

Description: 3–4mm, rusty red with up to 13 black spots on each elytron and with a coating of fine hair, although this may be visible only with a lens. No other British ladybird is hairy. The spot pattern is very variable and the spots are often joined together.

Food and habits: An herbivorous species, feeding on a wide range of wild and cultivated plants: a carnation pest in France. Winter is passed in low-growing vegetation, but the beetle remains active for much of the time.

Habitat and range: Mainly grassland and waste ground: sometimes at the base of garden walls. Most of Europe except the far north.

1

2/3

3

1 Wasp Beetle

Clytus arietis

Description: 7–15mm, with a very rounded black thorax and variable yellow banding on the elytra.
Food and habits: Feeds on pollen and nectar from flowers. An excellent wasp mimic, the female is often seen scampering over trees and fences with its antennae quivering as it searches for egg-laying sites. May–July. Despite its name and appearance, it is completely harmless. The larvae develop in dead hardwoods, including fence posts.
Habitat and range: Woods, hedgerows, gardens, and orchards. All Europe except the far north.
Similar species: Plagionotus arcuatus has yellow spots on the thorax. Probably extinct in the British Isles, it may breed in building timbers.

2 Colorado Beetle

Leptinotarsa decemlineata

Description: 10mm, with bold yellow and black stripes on the elytra.
Food and habits: A serious pest of potatoes. Both adults and the fleshy pink or orange larvae (2a) strip the leaves and reduce the plants to stumps, leading to drastic reductions in yield. Adults can be found throughout the year, although they are dormant in the soil in winter.
Habitat and range: Potato fields and gardens: occasionally on tomatoes and on nightshades in the surrounding countryside. This North American insect is now established in most parts of Europe, but not in the British Isles, where it should be reported to the police if found.

3 Asparagus Beetle

Crioceris asparagi

Description: 5–8mm, with a rust-coloured thorax and borders to the elytra. The 3 more or less square cream patches on each elytra vary in size and are sometimes separated by just narrow black lines.
Food and habits: Feeds on asparagus leaves and sometimes causes serious damage. The larvae, grey with black spots, feed on the same plants. May–August. Winter is passed as dormant adults in the soil.
Habitat and range: Southern and central Europe, including southern England.

4 Lily Beetle

Lilioceris lilii

Description: 6–8mm: brilliant red apart from its black head and legs.
Food and habits: Feeds on various wild and cultivated plants in the lily family, often completely stripping their leaves. April–August, with two or three broods in a year. The larvae are orange, but they coat themselves with their slimy black excreta and look more like slugs. Winter is passed in the adult stage, dormant in the soil.
Habitat and range: Mostly in parks and gardens. Most of Europe except the far north. Rare in the British Isles until recently, but now spreading rapidly as a result of increasing horticultural trade.

1

2/2a

3/4

1 Pea Beetle

Bruchus pisorum

Description: 4–5mm, dark brown with truncated elytra decorated with patches of white hair.

Food and habits: Adults feed on dried pea seeds. Larvae are the plump, white grubs found in pea pods, usually one inside each pea. Winter is passed as a dormant adult, still inside the seed.

Habitat and range: Wherever wild or cultivated peas grow. All Europe except the far north.

2 Pea Weevil

Sitona lineatus

Description: 4–5mm, with light and dark stripes on the elytra.

Food and habits: Adults, active mainly autumn and spring, chew semi-circular notches in the edges of leaves of peas, clovers, and other legumes and may cause severe damage to seedlings. Larvae, living inside the root nodules, also damage peas and other leguminous crops. Adults often swarm on sunny walls in autumn before overwintering in turf or leaf litter.

Habitat and range: Wherever wild or cultivated legumes grow. All Europe except the far north.

Similar species: There are many similar species.

3 Vine Weevil

Otiorhynchus sulcatus

Description: 8–12mm: black with a strongly domed and oval rear end. Long, elbowed antennae are attached to the upper surface of the broad, blunt snout. The elytra are fused together and the weevils cannot fly.

Food and habits: Adults nibble the leaves of a wide range of woody and herbaceous plants and also kill young twigs by nibbling away their bark. The legless white larvae destroy roots. Winter is usually passed in the pupal stage, but adults can be found throughout the year in greenhouses. Males are very rare in this species and reproduction is almost entirely parthenogenetic, with the females laying unfertilised but fertile eggs.

Habitat and range: Most common in parks and gardens, with the larvae particularly troublesome in greenhouses and on pot plants in the house. Out of doors in Southern and Central Europe: in doors elsewhere.

1

2

3

Butterflies

Butterflies and moths belong to the order Lepidoptera. The adults suck nectar, but they spend their early lives as leaf-chewing caterpillars. Butterflies fly by day and most moths are nocturnal, but the best way to tell them apart is to look at the antennae: butterfly antennae always have clubbed or swollen tips. Every species of butterfly has probably been seen in a garden somewhere or other, but only about 20 species can really be regarded as garden butterflies – and even fewer actually breed in gardens.

1 Scarce Swallowtail

Iphiclides podalirius

Description: Forewing 30–45mm. The ground colour is creamy white to pale yellow, with similar markings on both surfaces.
Flight: March–September in one or two broods, summer insects being paler than spring ones.
Caterpillar and life cycle: Caterpillar up to 40mm, is green and slug-shaped, with red spots and slender yellow stripes. Feeds on blackthorn and other trees, including cultivated fruit trees. Winter is passed as a chrysalis.
Habitat and range: Gardens, orchards, hedgerows, and scrubby places. Southern and central Europe, but absent from the British Isles. Despite its English name, it is the commonest swallowtail in many areas.

2 Black-veined White

Aporia crataegi

Description: Forewing 25–35mm. The black or dark brown veins are clearly visible on both surfaces of the thinly scaled wings.
Flight: May–July.
Caterpillar and life cycle: Caterpillars up to 40mm, are hairy, grey and black with orange or yellow spots. Live communally in silken tents for much of their lives. Feed on rosaceous trees and shrubs and can cause damage to fruit trees. Winter is passed as a small caterpillar in the tent.
Habitat and range: Open country with scattered trees or scrub. All Europe except far north and British Isles, where it became extinct in 1920s.

3 Brimstone

Gonepteryx rhamni

Description: Forewing 25–30mm. The male upperside is brilliant yellow, but only the paler, leaf-like underside can be seen when the butterfly settles. The female is greenish white and often mistaken for a Large White in flight, although she has no black markings.
Flight: June–October, and again in spring after hibernation: the adult lives for about a year and flies in all but the coldest months.
Caterpillar and life cycle: The bluish green caterpillar, up to 33mm, feeds on buckthorn and alder buckthorn and is unlikely to be seen in the garden. Winter is passed as a sleeping adult, usually in evergreen shrubs where the leaf-like wings afford excellent camouflage.
Habitat and range: Woodland margins, hedgerows, gardens, and other scrubby places. All Europe except the far north.

1

2/2a

3

1 Large White

Pieris brassicae

Description: Forewing 25–35mm. The ground colour of the upperside is chalky white, with black tips reaching at least half way down the edge of the forewing. The male has only one black spot on the forewing. Both sexes have two black spots on the underside. The underside of the hindwing is yellowish with a variable dusting of black scales.
Flight: April–October in two or three broods.
Caterpillar and life cycle: The caterpillars, up to 40mm, are yellowish green with black spots. They live gregariously on cabbages and other brassicas and also on garden nasturtiums, often reducing the leaves to skeletons. Winter is passed as a chrysalis. (See also p.144.)
Habitat and range: Gardens and other flowery places. Throughout Europe.
Similar species: Small White is smaller, with a much smaller black patch on the wing-tip. Both species are commonly called cabbage whites.

2 Small White

Pieris rapae

Description: Forewing 15–30mm. The upperside is white, with one black or grey spot on the forewing in the male and two in the female. There is a small black or grey patch at the wing-tip, extending further along the front margin than down the outer edge. Both sexes have two spots on the underside of the forewing. The underside of the hindwing is yellow, with a variable amount of grey dusting.
Flight: March–October in two or more broods.
Caterpillar and life cycle: The caterpillar, up to 25mm, is bluish green with a velvety texture and a yellow line on the back. It feeds singly on various brassicas, on which it is very difficult to spot, and also on nasturtiums. Winter is passed inn the chrysalis stage. Like the Large White, it is a serious pest.
Habitat and range: Gardens and other flowery places. Throughout Europe.
Similar species: Large White is bigger, with bigger wing-tip patches.

3 Green-veined White

Pieris napi

Description: Forewing 18–30mm. The uppersides are like those of the Small White, although males do not always have a black spot on the forewing. The underside of the hindwing is usually quite yellow and the veins are outlined with black and yellow scales, giving them a green appearance, although this is much reduced in summer insects.
Flight: March–November, in two, three, or even four broods: single-brooded in far north.
Caterpillar and life cycle: The caterpillar is like that of the Small White, except that it has no yellow stripe on the back. It feeds on a wide range of wild crucifers and is not a garden pest. Winter is passed in the chrysalis stage.
Habitat and range: Gardens, hedgerows, woodland margins, marshes, and other flowery places. Throughout Europe.

1/1a
2/2a
3

1 Orange-tip

Anthocaris cardamines

Description: Forewing 20–25mm. The male is easily identified by his orange wing-tips, although these are not always visible at rest. The female has no orange and is sometimes mistaken for one of the whites, but she has much more rounded wing-tips and, like the male, she has a mottled green and white pattern on the underside of her hindwing.
Flight: March–July.
Caterpillar and life cycle: The caterpillar, up to 30mm, is greyish green on the back and dark green below. It feeds on the developing fruits of cuckooflower, garlic mustard, and other crucifers, including garden honesty and sweet rocket. Winter is passed in the chrysalis stage.
Habitat and range: Damp woods, hedgerows, marshes, gardens, and other flowery places. All Europe except the far north.

2 Small Copper

Lycaena phlaeas

Description: Forewing 10–17mm. The upperside of the forewing is orange or copper-coloured and very shiny, with a brown margin and scattered black spots. The upperside of the hindwing is brown with an orange border and sometimes a few blue spots. The underside of the hindwing is usually dull brown.
Flight: February–November, with two, three, or even four broods. The butterfly flies close to the ground and likes to bask on bare earth.
Caterpillar and life cycle: The caterpillar is green and slug-like, often with pink stripes, and reaches a length of about 15mm. Winter is passed in the caterpillar stage.
Habitat and range: Heaths, rough grassland, waste ground, parks, and gardens. Throughout Europe.

3 Holly Blue

Celastrina argiolus

Description: Forewing 12–18mm. The male upperside is bright violet blue with narrow black margins. The female upperside (pictured here) is more of a sky blue, with broader black margins – especially in the summer brood, where the black extends almost to the middle of the wing. The undersides of both sexes are powdery blue with elongated black spots.
Flight: April–September in two broods, often flying around tall trees. It feeds mainly on honeydew and rarely visits flowers.
Caterpillar and life cycle: The caterpillar is slug-shaped and up to 12mm. Bright green with pink or purple stripes when young, it becomes brownish red when mature. Spring caterpillars feed on the flowers and developing fruits of holly and numerous other shrubs: late summer caterpillars feed on ivy flowers. Winter is passed as a chrysalis.
Habitat and range: Woodland margins, hedgerows, parks, and gardens. All Europe except Scotland and the far north. This is the only blue likely to be found in gardens in most parts of Europe.

1/1a
2
3

1 Large Tortoiseshell

Nymphalis polychloros

Description: Forewing 30mm. The upperside is essentially dull orange with black spots and a black border, the latter containing blue spots on the hindwing. The underside is mottled brown with a faint blue border.
Flight: June–August and again in spring after hibernation, feeding mainly on honeydew and on sap oozing from wounded trees.
Caterpillar and life cycle: The caterpillars, up to 45mm, are black with brownish spines and orange stripes. They feed mainly on elms and sallows, spending most of their lives in communal silk tents. Adults sleep through the winter, often in sheds and attics.
Habitat and range: Most of Europe but becoming rare in most areas: probably extinct in British Isles apart from casual immigrants.
Similar species: Small Tortoiseshell has basal half of hindwing black.

2 Small Tortoiseshell

Aglais urticae

Description: Forewing 25mm. The upperside is essentially bright orange with black borders containing blue spots. There are three rectangular black spots at the front of the forewing, and the basal half of the hindwing is black. The underside is mottled black and brown, with a row of faint blue spots near the outer edge.
Flight: On the wing almost throughout the year – May–October, in one, two, or three broods, and again in early spring after hibernation.
Caterpillar and life cycle: The caterpillar, up to 22mm, is brown or black with branching spines and two yellow bands on each side. It feeds communally on stinging nettles. Adults sleep through the winter, often in sheds and houses.
Habitat and range: Gardens and other flowery places throughout Europe. In a recent survey it was the commonest butterfly in every garden studied.
Similar species: Large Tortoiseshell is duller, with little black on the hindwing. Painted Lady is much paler, with black dots on the hindwing.

3 Painted Lady

Cynthia cardui

Description: Forewing 25–20mm. Upperside is pale orange with black markings, including a broad black triangle at the wing-tip containing white spots. Underside of the forewing is a paler version of the upperside, but underside of the hindwing is mottled brown and grey with a row of eye-spots near the edge.
Flight: April–October in two broods.
Caterpillar and life cycle: The caterpillar is largely black, with branched black or yellowish spines and a broken yellow line on each side. Up to 30mm, it feeds mainly on thistles, hiding between leaves spun together with silk. The species does not winter in Europe.
Habitat and range: Gardens and other flowery places. A summer visitor to all parts of Europe from Africa, sometimes in enormous numbers.
Similar species: Tortoiseshells lack the white spots near the wing-tip.

1

2/2a

3

1 Red Admiral

Vanessa atalanta

Description: Forewing 30mm. The velvety black upperside, boldly marked with red and white, makes this butterfly easy to identify. The underside is similar, although the mottled brown hindwing has no red margin.
Flight: May–October, usually in two broods, and again in spring after hibernation. Very fond of over-ripe fruit in the garden.
Caterpillar and life cycle: The caterpillar, up to 35mm, ranges from yellowish brown to black, with a row of yellow spots on each side. It has short black spines and is heavily speckled with white. It feeds on stinging nettles, where it hides between two leaves spun together with silk. Winter is passed as a sleeping adult, although few specimens survive the winter in the British Isles.
Habitat and range: Gardens and all other flowery places. Resident in the south and migrating to all parts of Europe in spring and summer.

2 Peacock

Inachis io

Description: Forewing 30mm. The four large eye-spots on the upperside make this butterfly quite unmistakable. The underside is almost black, providing excellent camouflage when the insect is hibernating.
Flight: June–September, and again in spring after hibernation: adult lives for about a year. Very fond of garden buddleia.
Caterpillar and life cycle: The caterpillar, up to 45mm, is black with tiny white spots and is clothed with branched spines. It feeds gregariously on stinging nettles. Adults sleep through the winter, often in sheds and attics.
Habitat and range: Gardens, parks, and all other flowery places. All Europe except the far north.

3 Comma

Polygonia c-album

Description: Forewing 23mm. The jagged wing margins immediately identify this butterfly. The ground colour of the upperside ranges from light to deep orange, while the underside ranges from light brown to a rich chocolate brown. The underside of the hindwing carries the white comma-shaped mark that gives the butterfly its name.
Flight: June–September, in two broods, and again in spring after hibernation. Fond of over-ripe fruit in the garden.
Caterpillar and life cycle: The caterpillar, up to 35mm, is brown with branched spines, and when mature it has a large white patch at the rear. It feeds on stinging nettles, elms, and hops. Winter is passed as a sleeping adult, mainly in shrubs and hedgerows, where the leaf-like undersides of the wings provide excellent camouflage.
Habitat and range: Woodland margins, hedgerows, parks, and gardens. Most of Europe, but absent from Ireland, Scotland, and the far north.
Similar species: Southern Comma of southern Europe, has smaller and fewer spots and the white mark on the underside is more like a 'V'.

1

2

2a/3

1 Ringlet

Aphantopus hyperantus

Description: Forewing 20mm. The cream-ringed eye-spots on the plain brown underside give this butterfly its name. The upperside is dark brown, usually with a few faint eye-spots.

Flight: June–August. Very fond of bramble blossom.

Caterpillar and life cycle: The caterpillar, about 20mm with a short forked 'tail', is pale brown with a dark line along the back. It feeds on various grasses, mainly at night. Winter is passed as a caterpillar.

Habitat and range: Woodland rides and clearings, and shady hedgerows, especially in damp places. Not uncommon in rural gardens with plenty of hedgerows in the vicinity. Most of Europe except the far north and south.

2 Gatekeeper

Pyronia tithonus

Description: Forewing 17–25mm. The uppersides are rich orange with broad brown borders and a large twin-pupilled eye-spot near the wing-tip. The male has a brown streak in the centre of the forewing. The underside of the forewing resembles the upperside, but the underside of the hindwing is largely brown, with yellow patches in the outer half.

Flight: July–September. Very fond of bramble blossom and marjoram.

Caterpillar and life cycle: The caterpillar is green or greyish brown with dark brown and cream stripes. It is up to 25mm, with a short forked 'tail'. It feeds on various grasses, usually at night. Winter is passed as a small caterpillar.

Habitat and range: Woodland margins, hedgerows, and scrubby grassland. Also known as the Hedge Brown, it is common in rural gardens with plenty of hedgerows in the vicinity, and also in orchards and shrubberies with long grass. Southern and Central Europe: absent from Scotland.

3 Wall Brown

Lasiommata megera

Description: Forewing 17–25mm. The uppersides are rich orange with brown borders and brown streaks of various intensity. There is a prominent eye-spot near the wing-tip and an arc of three or four eye-spots on the hindwing. The underside of the forewing resembles the upperside, but the underside of the hindwing is mottled brown and pearly grey, with six eye-spots and often with a strong golden sheen.

Flight: March–October in two or three broods. Fond of settling on rocks and walls.

Caterpillar and life cycle: The caterpillar, up to 25mm, is bluish green with faint white stripes. It is rather hairy and has two short 'tails'. It feeds on various grasses, mainly at night. Winter is passed as a caterpillar.

Habitat and range: Rough grassland, including roadside verges from which it often enters gardens. May breed in grassy orchards and shrubberies. Most of Europe except northern Scandinavia.

1

2

3

Moths

Moths are mainly nocturnal insects. They are often seen flying away when disturbed amongst the flowers or vegetables during the daytime, but otherwise they are most likely to be seen resting on tree trunks and fences by day, although many of them are beautifully camouflaged and not easy to spot. Although closely related to butterflies, their antennae do not normally have swollen tips: most are either hair-like or feathery. Many moths breed in our gardens and quite a few of their caterpillars damage our flowers and crops. Adults of many species are attracted to lighted windows at night.

1 Death's Head Hawkmoth

Acherontia atropos

Description: Forewing up to 60mm. The size and colour, together with the skull-like pattern on the thorax, immediately identify this moth.

Flight: April–November. Rarely visits flowers, but drinks sap oozing from trees and also enters bee-hives for honey. The moth squeaks when handled, by forcing air through its short tongue.

Caterpillar and life cycle: Up to 125mm, the caterpillar is usually yellow with seven blue or purple diagonal stripes, and a curly horn at the rear. There is also a brown form, in which the first three segments are white and resemble a little shawl. The caterpillar feeds on potatoes and various nightshades, and sometimes eats privet. Winter is passed in the chrysalis stage. Chrysalises are occasionally dug up with potatoes in autumn, but they cannot normally survive the British winter.

Habitat and range: Found in many habitats, but most often seen in and around potato fields and gardens. A summer visitor from Africa, reaching most parts of Europe although never common. Most specimens are seen in the south, where a few pupae may survive the winter.

2 Privet Hawkmoth

Sphinx ligustri

Description: Forewing up to 55mm, largely brown with black streaks and heavily dusted with white or pink at the front. Thorax is black with a white streak at each side; abdomen and hindwings have pink and black bands. At rest, the wings are pulled tightly back and held tent-like over the body. The insect then resembles a broken twig. The largest resident British moth.

Flight: June–July, feeding on the wing by plunging its long tongue into various flowers, including garden honeysuckle.

Caterpillar and life cycle: Up to 100mm, the caterpillar is bright green with purple and white diagonal stripes, and a shiny black horn at the rear. Despite its size, it is surprisingly well camouflaged amongst the foliage. It feeds mainly on privet, but also eats ash and lilac and sometimes elder. Winter is passed as a chrysalis deep in the ground.

Habitat and range: Woodland margins, hedgerows, parks, and gardens – including suburban areas where many gardens are surrounded by privet hedges. Most of Europe except Ireland, Scotland, and the far north.

Similar species: Convolvulus Hawkmoth (p.100) has greyer forewings and a grey thorax, and has no pink on the hindwings.

1

1a/2

2a

1 Eyed Hawkmoth

Smerinthus ocellata

Description: Forewing up to 45mm, with a wavy outer edge: a mixture of light and dark brown, strongly tinged with pink in the basal half. The thorax is greyish brown with a broad chocolate brown stripe in the middle. The hindwing is pink and brown with a large eye-spot at the rear. The eye-spots are concealed by the forewings when the moth is at rest, but the moth raises its forewings when disturbed and the sudden appearance of two large 'eyes' is enough to scare off most predators.
Flight: May–September in two broods. It does not feed.
Caterpillar and life cycle: The caterpillar, up to 70mm, is leaf-green with seven diagonal yellow stripes and numerous pale spots. It has a bluish horn at the rear. It feeds mainly on willows and apple, usually upside-down and very well camouflaged. Winter is passed as a chrysalis in the ground.
Habitat and range: Light woodland, parks, and gardens. Throughout Europe except Scotland and the far north.

2 Poplar Hawkmoth

Laothoe populi

Description: Forewing about 40mm, ranging from ash grey to pinkish brown with a darker central band. Hindwing has an orange patch at the rear. Both wings have scalloped outer edges and when the moth is at rest, with its hindwings projecting in front of the forewings, it is easily mistaken for a cluster of dead leaves. It deters predators by displaying the orange spots when disturbed (see above).
Flight: May–September, in two broods. The moth does not feed.
Caterpillar and life cycle: The caterpillar, up to 60mm, is bright green or bluish green with seven diagonal yellow stripes and numerous yellow spots. It has a yellow horn, sometimes tipped with red. It feeds on poplars and sallows. Winter is passed as a chrysalis in the ground.
Habitat and range: Woodland margins, river valleys, and wherever else the food-plants grow, including urban parks and gardens. Throughout Europe except the far north.

3 Lime Hawkmoth

Mimas tiliae

Description: Forewing about 35mm with a 'ragged' outer margin. It ranges from green to pink or orange and usually has a dark green band in the middle. At rest, the moth is easily mistaken for a dead leaf.
Flight: May–July. It does not feed.
Caterpillar and life cycle: The caterpillar, up to 60mm, is leaf-green with diagonal yellow stripes and pale spots. It tapers strongly towards the front and the horn is largely blue. It feeds on lime, elm, birch, and several other deciduous trees. Winter is passed as a chrysalis in the ground.
Habitat and range: Woods, parks, large gardens, and tree-lined avenues. Most of Europe except Scotland, Ireland, and the far north.

1

2/2a

3

1 Convolvulus Hawkmoth

Agrius convolvuli

Description: Forewing up to 55mm, grey with a few black streaks. The thorax is grey and the abdomen is grey with pink bands. Hindwing is grey with darker bands.

Flight: June–November. Feeds on the wing, plunging its enormously long tongue – much longer than the body – into various tubular flowers, including honeysuckle and tobacco.

Caterpillar and life cycle: The caterpillar, about 100mm, is either apple-green with seven pale brown diagonal stripes or dark brown with yellow or pink stripes. Both forms have a black-tipped reddish brown horn. The caterpillars feed on bindweeds and morning glories.

Habitat and range: Occurs in a wide range of open and scrubby habitats, including urban parks and gardens. A visitor from Africa, spreading to all parts of Europe during the summer and breeding in many places, mainly in the south: unlikely to survive the winter anywhere in Europe.

Similar species: Privet Hawkmoth (p.96) is browner and has pink-striped hindwings.

2 Elephant Hawkmoth

Deilephila elpenor

Description: Forewing up to 35mm: bronzy green with pink stripes. Hindwing is bright pink with a black base.

Flight: May–July, sometimes with a small second brood later. Feeds on the wing at honeysuckle and other tubular flowers.

Caterpillar and life cycle: The caterpillar is usually dark brown, but may be green, with two pairs of pink and black eye-spots just behind the trunk-like snout that gives the insect its name. When alarmed, the snout is withdrawn into the eye-spot region, which swells up and sways menacingly from side to side. Up to 90mm, the caterpillar has a small horn and feeds on willowherbs, bedstraws, and garden fuchsias.

Habitat and range: Woodland margins, parks, and gardens, and open country of all kinds. Throughout Europe.

3 Hummingbird Hawkmoth

Macroglossum stellatarum

Description: Forewing about 25mm and dull brown. Hindwing is orange, although all that one normally sees of this day-flying moth is a brownish blur as it hovers in front of flowers and probes them with its long tongue. It makes an audible hum like a humming-bird.

Flight: Day-flying, throughout the year in the south, but usually April–September in the British Isles. There are two or more broods each year.

Caterpillar and life cycle: The caterpillar, up to 60mm, is green or brown with pale lines on the sides and a yellow-tipped blue horn at the rear. It feeds on bedstraws. Winter is passed in the adult state, although it is rare for moths to survive the winter north of the Alps.

Habitat and range: Parks, gardens, and other flowery habitats. Resident in southern Europe, spreading northwards to most parts in the summer.

1

2

2a/3

1 Buff-tip Moth

Phalera bucephala

Description: Forewing up to 30mm, silvery grey with a more or less circular buff tip. At rest, the wings are wrapped tightly around the body and, together with the buff hairs of the thorax, they give the insect a striking resemblance to a broken twig.
Flight: May–August.
Caterpillar and life cycle: The caterpillar, up to 80mm, is deep yellow with broken black stripes running the length of the body. It feeds gregariously for much of its life, stripping the leaves from whole branches of oaks, limes, elms, and many other deciduous trees. Winter is passed as a chrysalis in the soil.
Habitat and range: Almost anywhere with trees and shrubs, including orchards and garden hedges. Throughout Europe.

2 Puss Moth

Cerura vinula

Description: Forewing about 35mm: white or pale grey and crossed by several zig-zag grey lines in the outer region. Very hairy.
Flight: April–July.
Caterpillar and life cycle: The caterpillar, about 70mm when mature, is bright green with a dark brown or purplish 'saddle'. When alarmed, it raises its front end and pulls its head into its thorax, producing a sinister 'face'. At the same time, it waves red filaments extruded from its much-modified hind legs. It feeds on willows and poplars and several other deciduous trees. Winter is passed in a tough cocoon, usually spun in a bark crevice.
Habitat and range: Woods and parks and wherever else its food-plants grow. Throughout Europe.
Similar species: Cerura erminea, not found in the British Isles, has conspicuous black loops at the base of the forewing.

3 Pale Tussock

Calliteara pudibunda

Description: Forewing 20–35mm: pale grey with variable darker cross-lines. Female noticeably larger and paler than male, pictured here. Very hairy.
Flight: April–July. The moth does not feed.
Caterpillar and life cycle: The unmistakable caterpillar, up to 45mm, is bright green or yellow with four dense white tufts like tiny shaving brushes on its back. There is also a slimmer tuft of red hairs at the back. The hairs can cause a rash if the caterpillar is handled. It feeds on a wide range of deciduous trees and also on hops. Winter is passed as a chrysalis.
Habitat and range: Woods, hedgerows, parks, and gardens. Most of central and northern Europe, but absent from Scotland.

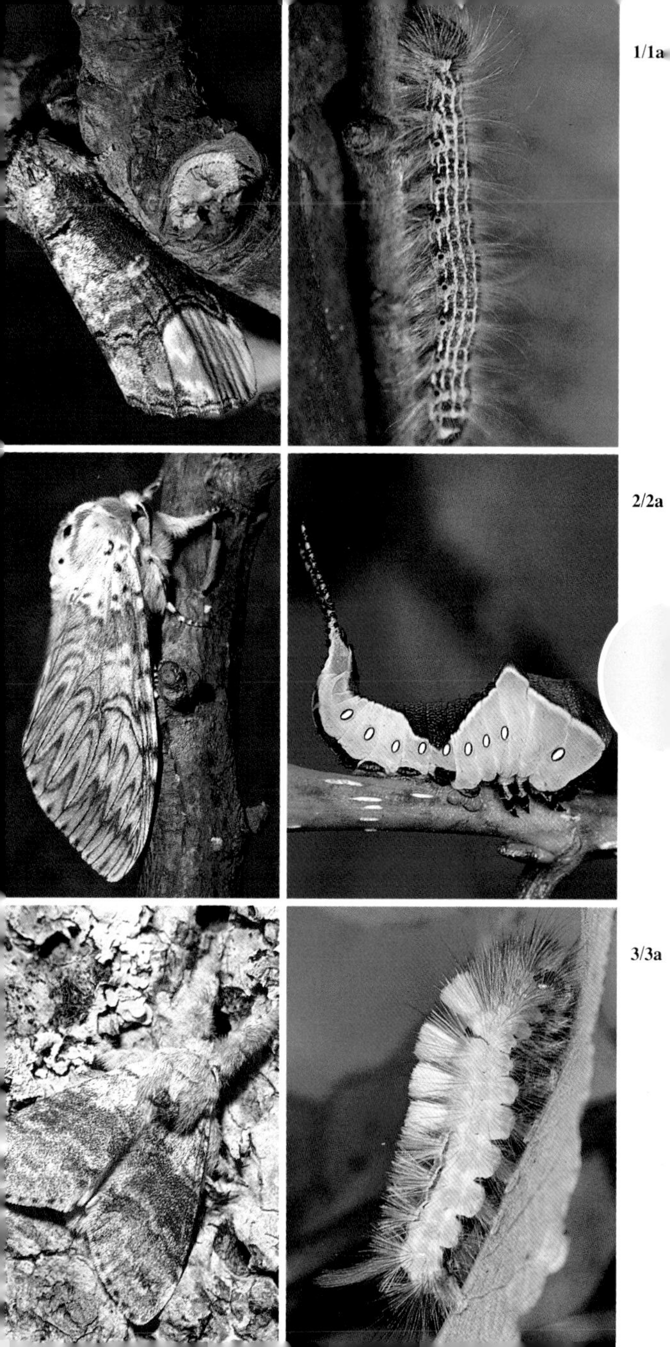

1/1a
2/2a
3/3a

1 Vapourer Moth

Orgyia antiqua

Description: Male forewing 15–18mm: rich chestnut, with black lines and a conspicuous, comma-shaped white spot at the rear. Female is wingless and never moves far from her cocoon.

Flight: June–October in two or three broods. The male usually flies by day and does not feed.

Caterpillar and life cycle: The caterpillar, up to 25mm, is dark grey with red spots and four tufts of cream or light brown hairs near the middle. There are two dark, horn-like tufts at the front and a similar tuft projects back from the rear of the body. The hairs can cause serious skin irritation if the caterpillar is handled. It feeds on a wide range of deciduous trees and shrubs and often damages orchard trees. Winter is passed in the egg stage, the eggs being laid in a large cluster on or close to the female's cocoon.

Habitat and range: Almost anywhere with trees and shrubs, including town parks and tree-lined streets. Throughout Europe.

2 Yellow-tail

Euproctis similis

Description: Forewing up to 20mm. Both wings are silky white, usually with a dark spot near the rear corner of the male forewing. The abdomen ends in a tuft of golden hairs. When disturbed, the moth often rolls over and pretends to be dead, as pictured here.

Flight: June–August. It does not feed.

Caterpillar and life cycle: The caterpillar is black with red stripes and white spots. It is up to 40mm and very hairy and can cause severe skin irritation if handled. It feeds on hawthorn and many other deciduous trees and shrubs. Winter is passed as a small caterpillar.

Habitat and range: Almost anywhere with trees and shrubs: often common in tree-lined streets. Most of Europe, but rare in Ireland and Scotland.

Similar species: The Brown-tail Moth has a brown tuft at the rear. Its caterpillars live communally and often damage orchard trees.

3 Lackey Moth

Malacosoma neustria

Description: Forewing up to 20mm: buff to brick-red with two light or dark cross-lines, the area between them often darker than rest of wing.

Flight: June–August.

Caterpillar and life cycle: The caterpillar is up to 55mm – rather long for the size of the adult – and basically bluish grey with longitudinal red stripes. The head has two conspicuous black spots. The caterpillars live in communal webs for most of their lives, feeding on the leaves of hawthorn, blackthorn, and many other trees and often damaging cultivated plums and other orchard trees. Winter is passed in the egg stage, in neat collar-like batches surrounding the twigs.

Habitat and range: Woods, hedgerows, parks, gardens, and tree-lined streets. Most of Europe except Scotland and the far north.

1/1a

2/2a

3/3a

1 Garden Tiger Moth

Arctia caja

Description: Forewing 25–35mm, essentially brown with a very variable cream network that provides excellent camouflage when the moth is at rest. The hindwing is deep orange or red with blue-black spots and is exposed as a warning whenever the moth is disturbed. The thorax is clothed with dense brown hair and has a narrow red collar.
Flight: June–August. It does not feed.
Caterpillar and life cycle: The caterpillar, often called a woolly bear because of its long black and chestnut hair, is up to 60mm. It feeds on a wide range of herbaceous plants, including many garden flowers and vegetables, but is most often seen scurrying across paths when looking for somewhere to pupate in the summer. Winter is passed as a small caterpillar.
Habitat and range: Almost any open habitat, including gardens and waste ground. Throughout Europe.

2 White Ermine

Spilosoma lubricipeda

Description: Forewing 15–20mm: chalky white with very variable black spots. The abdomen is yellow with black spots.
Flight: May–August. It does not feed.
Caterpillar and life cycle: The caterpillar, up to 45mm, has a shiny black head and its body is clothed with tufts of dark brown or black hair. There is a dark red or orange line along the back. It feeds on a wide range of herbaceous plants, including docks and dandelions as well as many cultivated species. Winter is passed as a chrysalis.
Habitat and range: Hedgerows, gardens, waste ground, and many other habitats. Throughout Europe.
Similar species: The female Muslin Ermine has rather silky white wings, usually with fewer spots, and a white abdomen.

3 Buff Ermine

Spilosoma lutea

Description: Forewing 15–20mm: pale cream to yellow, with variable dark markings including a broken line running diagonally back from the wing-tip. The male, pictured here, is usually darker than the female.
Flight: May–August.
Caterpillar and life cycle: The caterpillar, up to 45mm, has a shiny brown head and its body is clothed with tufts of brown hair. It feeds on a wide range of both wild and cultivated herbaceous plants. Winter is passed as a chrysalis.
Habitat and range: Almost any habitat, but especially common in hedgerows and gardens and on waste ground. Throughout Europe.

1/1a

2/2a

3/3a

1 Grey Dagger

Acronicta psi

Description: Forewing 15–20mm: pale to dark grey with a number of black crosses or dagger-like marks.

Flight: May–September in one or two broods: very well camouflaged on lichen-covered fences and tree trunks.

Caterpillar and life cycle: The caterpillar, up to 40mm, is easily recognised by its pale yellow back with a black horn on the 1st abdominal segment. Each side has a dark grey band enclosing red spots, with a white stripe below it. It feeds on hawthorn and many other deciduous trees and shrubs. Winter is passed as a chrysalis.

Habitat and range: More or less anywhere with trees and shrubs, including town parks and gardens. Throughout Europe except the far north.

Similar species: Dark Dagger is almost identical, but its caterpillar has a white back and only a small hump on 1st abdominal segment.

2 Cabbage Moth

Mamestra brassicae

Description: Forewing 15–25mm: mottled greyish brown with a variable amount of rust-coloured scales and usually with a prominent, white-edged kidney shaped mark just beyond the middle. A pale wavy line is commonly present near the outer edge of the wing, although it is often weak or absent in specimens from the north. The hindwing is silvery grey with a darker margin.

Flight: Mainly May–September, although it can be seen on the wing at any time of year.

Caterpillar and life cycle: The plump, fleshy caterpillar is up to 50mm and is usually brownish green when mature, but it is sometimes bright green or flesh-coloured. There is a row of short black dashes on each side. It feeds on a wide range of herbaceous plants, often boring into cabbages and contaminating them with its strong-smelling droppings. Winter is passed as a chrysalis in the soil.

Habitat and range: Almost any habitat, but particularly common in cultivated areas. Throughout Europe except the far north.

3 Dot Moth

Melanchra persicariae

Description: Forewing up to 20mm: mottled bluish black with a prominent white kidney-shaped spot just beyond the middle.

Flight: May–September.

Caterpillar and life cycle: The caterpillar is up to 45mm and ranges from greyish green to brown or purple, with conspicuous light or dark diagonal stripes on the back and the sides. It feeds on a wide range of herbaceous plants and low-growing shrubs, but is especially fond of stinging nettles. Winter is passed as a chrysalis in the ground.

Habitat and range: Found in a wide range of habitats, but most common in and around human settlements. Most of Europe, but absent from the far north and most of Scotland.

1/1a
2
3/3a

1 Large Yellow Underwing

Noctua pronuba

Description: Forewing about 25mm: light greyish brown to rich chestnut, often marbled and with two pale spots or rings near the middle. Hindwing is deep yellow with a black outer border and flashes conspicuously in flight when the moth is disturbed. Flight is fast and erratic, but the moth quickly lands again and 'disappears'. The wings are laid flat over the body at rest.
Flight: June–October.
Caterpillar and life cycle: The plump, fleshy caterpillar, up to 50mm, is green or brown with two rows of black dashes on the back. It hides in the soil by day and feeds at night, remaining active all through the winter. It attacks many garden plants, often cutting the stems at ground level – and leaving slugs to take the blame. Several other caterpillars feed in this way and are known as cutworms.
Habitat and range: Almost any well-vegetated habitat. Throughout Europe except the far north.
Similar species: The Broad-bordered Yellow Underwing has a very broad black border to the hindwing and wraps its wings slightly round its body at rest. There are several smaller species.

2 Heart and Dart

Agrotis exclamationis

Description: Forewing about 20mm: pale greyish brown to deep brown, usually with two conspicuous, dark marks that give the moth its name.
Flight: May–September in one or two broods: autumn moths are usually smaller.
Caterpillar and life cycle: The caterpillar, about 40mm, is dull brown above and grey below. It is another one of the cutworms (see above) and feeds on almost any wild or cultivated herbaceous plant. Winter is passed as a dormant, fully-grown caterpillar.
Habitat and range: Almost anywhere, but especially common on cultivated land. Throughout Europe except the far north.

3 Setaceous Hebrew Character

Xestia c-nigrum

Description: Forewing up to 20mm: greyish brown to chestnut, often tinged with purple and bearing a conspicuous black mark shaped like a bow-tie. A pale patch extends from this mark to the front edge of the wing. The wings are laid flat at rest.
Flight: May–October, but most common in autumn.
Caterpillar and life cycle: The young caterpillar is green, but becomes greenish or greyish brown with black, wedge-shaped streaks on the back. It is about 40mm when mature. It eats a wide range of herbaceous plants. Winter is passed as either a caterpillar or a chrysalis.
Habitat and range: Almost anywhere. All Europe except the far north.
Similar species: Hebrew Character has a less obvious pale patch on the front edge of the wing and holds its wings tent-like at rest.

1

2

3

1 Angle Shades

Phlogophora meticulosa

Description: Forewing about 25mm, with a somewhat ragged and distinctly concave outer margin. The ground colour ranges from olive green, through pale brown, to pink and there is a V-shaped green or brown band across the middle, but the moth is easily recognised by its characteristic resting position, with the front edge of the forewing rolled or wrinkled. The resting moth is easily mistaken for a dead leaf.

Flight: All year, but most abundant in autumn.

Caterpillar and life cycle: The caterpillar, up to 45mm, is bright green or brown with a broken pale line on the back (often very faint) and a row of dark diagonal streaks on each side of it. There is also a white line on each side and a slight hump at the rear. It feeds on a wide range of shrubs and herbaceous plants and can be a pest in vegetable plots and herbaceous borders. Winter is passed mainly in the caterpillar stage.

Habitat and range: Virtually any habitat. The moth is a great migrant and occurs all over Europe, but only as a summer visitor in the far north.

2 Old Lady

Mormo maura

Description: Forewing 30–35mm, with a strongly scalloped outer margin. The moth's name is derived from the mottled and banded brown pattern, reminiscent of the shawls once worn by elderly ladies.

Flight: June–August: often rests in sheds and other out-buildings in the day-time. Feeds mainly on honeydew and sap oozing from damaged trees.

Caterpillar and life cycle: The rather flabby caterpillar, up to 75mm, is greyish brown with a dotted white line running between dark, diamond-shaped smudges along the back. The sides of each segment carry an oblique black stripe. The caterpillar feeds on various herbaceous plants in autumn, but turns to blackthorn and other shrubs when it wakes from hibernation in the spring.

Habitat and range: Woods, hedges, waste ground, and gardens: not uncommon in towns. Southern and Central Europe: rare in Scotland and Ireland.

3 Herald

Scoliopteryx libatrix

Description: Forewing 20–25mm, with a jagged outer margin: purplish brown, heavily dusted with bright orange scales at the base and crossed by a double white line. The moth commonly rests in a head-down position and then bears a strong resemblance to a dead leaf.

Flight: May–November, in one or two broods, and again in spring after hibernation.

Caterpillar and life cycle: The slender caterpillar, up to 55mm, is bright velvety green with a narrow yellow stripe on each side. It feeds on willows and poplars. Winter is passed as a sleeping adult, often in sheds and other buildings.

Habitat and range: Parks, gardens, and wherever else the food-plants grow. Most of Europe except the far north.

1/1a

2
3

1 Mullein Moth

Cucullia verbasci

Description: Forewing 20–25mm: straw-coloured to mid-brown, with darker streaks. The wings are pulled tightly back around the body at rest and the moth looks remarkably like a broken twig or sliver of bark.
Flight: April–June. Because of its excellent camouflage, the moth is rarely seen, but its caterpillar is only too well known to gardeners.
Caterpillar and life cycle: The caterpillar, up to 60mm, is creamy white with black and yellow spots. These warning colours allow it to feed openly on the leaves and flower spikes of mulleins, which it often reduces to blackened stumps. It also feeds on figworts and sometimes on buddleia. Winter is passed as a chrysalis in a thick underground cocoon, and the insects sometimes remain in this stage for two or three years.
Habitat and range: Woodland margins, river banks, and other scrubby places, as well as parks and gardens. Most of Europe except Scotland, Ireland, and the far north.

2 Burnished Brass

Diachrisia chrysitis

Description: Forewing up to 20mm, with two, broad shiny bands ranging from brassy green to deep gold. The brown band between them is often broken in the middle, allowing the two brassy or golden areas to merge.
Flight: May–October in one, two, or three broods: feeds on the wing.
Caterpillar and life cycle: The caterpillar, up to 35mm, is light bluish green with diagonal white streaks on the back and a white line on each side. Unlike most other caterpillars, it has only three pairs of abdominal legs, including the claspers at the rear. It feeds mainly on stinging nettles, but also on deadnettles and garden mint. Winter is passed as a small caterpillar.
Habitat and range: Hedges, waste ground, parks, and gardens. All Europe.

3 Golden Plusia

Polychrisia moneta

Description: Forewing up to 20mm: pale gold with a dark brown cross-line and a silver-edged figure-of-8 near the middle.
Flight: May–September in one or two broods.
Caterpillar and life cycle: The caterpillar is grey with black spots at first, but later becomes bright green with a white line on each side. Up to 40mm, it tapers strongly to the front and has only three pairs of abdominal legs – including the claspers. It eats a wide range of plants, but mainly garden delphiniums in the British Isles – nibbling the base of the leaf blade which then collapses and forms a shelter around the caterpillar. The chrysalis, wrapped in a thin silvery or golden cocoon, is half brown and half green. Winter is passed as a caterpillar in the soil.
Habitat and range: In the British Isles the moth is rarely found away from parks and gardens, but elsewhere it thrives equally well in uncultivated areas. Throughout Europe.

1/1a

2

3/3a

1 Silver-Y

Autographa gamma

Description: Forewing about 20mm, ranging from purplish grey to velvety black and often tinged with brown: There is a bright silvery Y in the middle, although this may be broken and the two arms may be indistinct.
Flight: Generally May–November, with two or more broods, but flies all year in southern Europe. It flies by day as well as at night and is most often seen as a grey blur hovering in front of flowers.
Caterpillar and life cycle: The caterpillar is up to 25mm and has only three pairs of fleshy abdominal legs, including the claspers. It varies from light bluish green to very dark green, with darker stripes, and rests with the front part of the abdomen raised, as shown here. It feeds on a very wide range of plants and causes damage to various crops when populations are high. The species breeds throughout the year in the Mediterranean area, but cannot survive the winter frosts further north.
Habitat and range: Almost any habitat. It is a great migrant and spreads to all parts of Europe in the summer, sometimes in immense numbers.
Similar species: Beautiful Golden Y and Plain Golden Y are much browner.

2 Chinese Character

Cilix glaucata

Description: Forewing 10–12mm, silky white with a brownish central blotch marked with silvery streaks. The outer region is smudged with grey and when at rest, with its wings held steeply tent-like above the body, the moth looks very much like a bird dropping.
Flight: May–September in two broods: often attracted to lighted windows.
Caterpillar and life cycle: The caterpillar, about 15mm, is reddish brown, with two 'warts' on the thorax and a rear end that tapers to a sharp point. It feeds on blackthorn and other rosaceous trees, including apples. Winter is passed as a chrysalis.
Habitat and range: Hedgerows and other shrubby places. Most of Europe except the far north.

3 Brindled Beauty

Lycia hirtaria

Description: Forewing 20–25mm. The ground colour of this rather hairy moth is pale brown or white, heavily speckled with dark brown and yellow and bearing several dark cross-lines. Very dark melanic forms (see p.118) are not uncommon in some urban areas.
Flight: February–May.
Caterpillar and life cycle: The caterpillar, up to 55mm, ranges from dirty green to purplish brown or chestnut, with yellow spots and a yellow band just behind the head. It is a typical looper (see p.118), with just two pairs of abdominal legs. It feeds on a very wide range of trees and shrubs. Winter is spent as a chrysalis in the soil.
Habitat and range: Woods, hedgerows, parks, gardens, and tree-lined urban streets. Most of Europe.

1/1a

2

3

1 Peppered Moth

Biston betularia

Description: Forewing 20–30mm. This moth occurs in two major forms, both pictured here. One, usually referred to as the normal form, is white with a dense speckling of black that gives the moth its name. The other is sooty black and is known as the melanic form. Only the normal form was known until the middle of the 19th century, although the melanic form undoubtedly occurred in very small numbers. As the industrial revolution got under way, the black form began to spread – because it was not so easily seen by birds when resting on smoke-blackened trees and walls. The normal form declined in industrial areas, where it was very conspicuous, and almost disappeared from industrial regions in the first half of the 20th century, although it remained common in many other places. Several other moths have evolved dark, melanic forms in the last hundred years or so and the phenomenon is known as industrial melanism. Melanic peppered moths are believed to be hardier or more tolerant of low-level pollution than the normal form, and they now occur in many rural areas as well as in industrial regions. But stricter pollution controls are now beginning to reverse the trend, and the normal form is increasing again, even in urban areas.

Flight: May–August.

Caterpillar and life cycle: The caterpillar, up to 60mm, is green or brown with a deeply notched head and is remarkably twig-like. It is one of a large group called loopers. It has only two pairs of abdominal legs and moves by stretching out and gripping a twig with its front legs before arching its body and drawing the rear end up to meet the front (see Magpie caterpillar below). It feeds on a wide range of shrubs, including raspberries and many other garden plants. Winter is spent as a chrysalis in the soil.

Habitat and range: Woods, hedgerows, parks, gardens, and all kinds of shrubby places. All Europe except the far north.

2 Magpie Moth

Abraxas grossulariata

Description: Forewing about 20mm. The pattern varies a good deal, but the sooty brown spots and the yellow line through the forewing readily identify this common moth. The bold pattern warns of an unpleasant taste and the moth is rejected by many birds and also by spiders.

Flight: June–August.

Caterpillar and life cycle: The caterpillar, about 30mm, has black spots on a white background and an orange or rust-coloured line along each side. It is a looper (see above), with only two pairs of abdominal legs. It feeds on blackthorn and a wide range of other deciduous shrubs and often causes damage to gooseberry and currant bushes in the garden, where it is sometimes mistaken for the larvae of the gooseberry sawfly (p.140). The black and yellow pupae are often found in flimsy cocoons on the leaves and twigs. Winter is passed as a small caterpillar.

Habitat and range: Woods, hedgerows, and gardens. Most of Europe except the far north.

1

1a

2/2a

1 Brimstone Moth

Opisthograptis luteolata

Description: Forewing 15–20mm. The brown flecks at the front of the bright yellow wings readily identify this moth.
Flight: April–October, in one or two broods: commonly attracted to lighted windows.
Caterpillar and life cycle: The caterpillar, up to 35mm, is some shade of brown, often heavily tinged with grey or green, and has a marked twin-pointed hump near the middle. A looper with only two pairs of abdominal legs, it is extremely twig-like and not easy to spot. It feeds on hawthorn and blackthorn and other rosaceous trees and shrubs, including cultivated apples. Winter is passed as a chrysalis.
Habitat and range: Woods, hedgerows, and gardens. Most of Europe.

2 Swallowtailed Moth

Ourapteryx sambucaria

Description: Forewing 25–30mm. An unmistakable moth, with a short, pointed 'tail' on each hindwing. The wings are bright lemon yellow at first, but soon fade to pale cream or almost white.
Flight: June–August: often attracted to lighted windows.
Caterpillar and life cycle: The slender caterpillar, up to 50mm, is an extremely twig-like looper, with just two pairs of abdominal legs. It is brown or dirty green and tapers strongly towards the head. It feeds on a wide range of trees and shrubs, especially hawthorn, blackthorn, and ivy. Winter is spent as a resting caterpillar, hiding in bark crevices or among the densely-packed twigs of the food-plant.
Habitat and range: Woods, hedgerows, gardens, and almost any other shrubby habitat. Most of Europe except the far north.

3 Garden Carpet

Xanthorhoe fluctuata

Description: Forewing 12–15mm, with a ground colour ranging from dirty white to dark grey. The darker markings range from sooty brown to black and vary in extent, although there is always a cloak-like patch around the 'shoulders' and a more or less rectangular patch near the middle of the wing. This patch sometimes extends back across the width of the wing, but the rear half is always paler than the rest. Like most other carpet moths, named for the intricate patterns of many species, the moth has a triangular outline at rest, with the hindwings concealed.
Flight: April–October, in two or three broods.
Caterpillar and life cycle: The caterpillar, about 25mm, ranges from green or grey to dark brown, with darker blotches on the back and often with a pinkish tinge below. A typical looper, it has only two pairs of abdominal legs. It feeds on wild and cultivated crucifers, including cabbages and wallflowers. Winter is passed as a chrysalis.
Habitat and range: Almost any habitat, but especially common in cultivated areas and on waste ground. Throughout Europe.

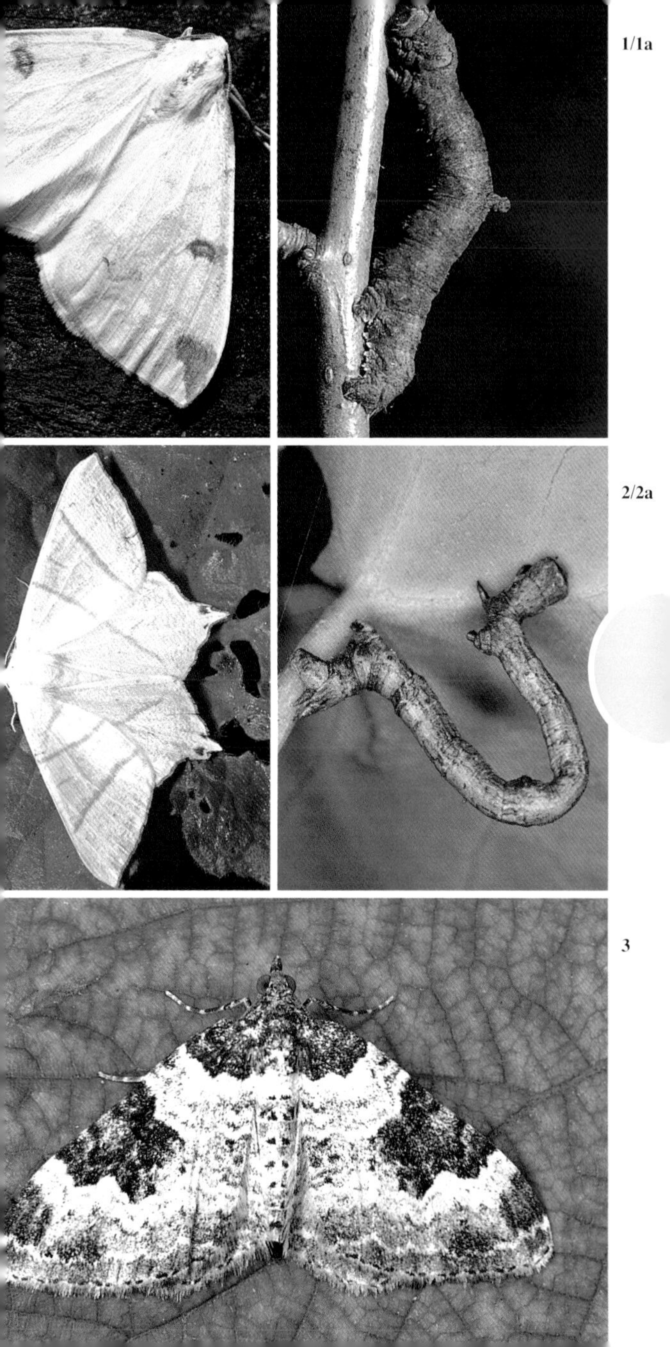

1/1a

2/2a

3

1 Bloodvein

Timandra griseata

Description: Forewing about 15mm. Ground colour ranges from cream to pale grey or buff, with pink borders and a pink or purple 'vein' running diagonally across each wing.
Flight: May–September in one or two broods: most common June–July.
Caterpillar and life cycle: The caterpillar, about 20mm, ranges from dark grey to chestnut, with diagonal light and dark stripes. At rest, the body often swells up just behind the head. It feeds on docks and other low-growing plants. Winter is passed as a caterpillar.
Habitat and range: Hedgerows and other rough habitats, especially where slightly damp. Most of Europe except the far north.

2 Mottled Umber

Erannis defoliaria

Description: Forewing of male about 20mm; ground colour usually ranges from off-white to brick-red, heavily dusted with darker scales. A dark, wavy band usually crosses the outer part of the wing, but many are almost plain brown with little trace of cross-lines. Outer margin of forewing always has small, dark dots. Female, illustrated below is wingless and spider-like.
Flight: September–March. Males often come to lighted windows: females sit on tree trunks and branches.
Caterpillar and life cycle: The caterpillar, a looper about 30mm, ranges from straw-coloured to rust-red, with a narrow black stripe and often rust-coloured blotches on each side. It feeds on almost any deciduous tree or shrub and can be a nuisance in orchards.
Habitat and range: Anywhere with trees and shrubs: very common in parks and gardens. All Europe except the far north.
Similar species: Dotted Border also has dark dots around the hindwing.

3 Winter Moth

Operophtera brumata

Description: Forewing of male about 15mm, ranging from pale grey to sooty brown with indistinct dots and cross-lines. Hindwings are greyish brown. Female wings are reduced to tiny stumps.
Flight: October–February. Males commonly come to lighted windows, even in winter. Females (illustrated below) sit on tree trunks and branches.
Caterpillar and life cycle: The caterpillar, a looper about 20mm, is green, with pale lines on the sides and a darker one on the back. Feeds on many deciduous trees and shrubs and is a serious pest of apples and pears, where it feeds on the blossom and leads to deformed fruit or no fruit at all.
Habitat and range: Almost anywhere with trees and shrubs. All Europe.
Similar species: Northern Winter Moth has whiter hindwings.

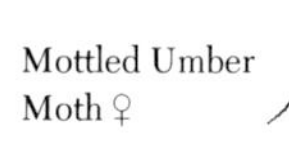
Mottled Umber Moth ♀

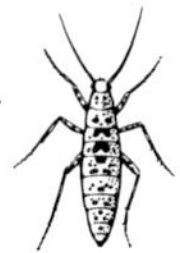

Winter Moth ♀

1

2/2a

3

1 Yellow Shell

Camptogramma bilineata

Description: Forewing 12–15mm. The ground colour is usually some shade of yellow, often heavily dusted with brown in northern and western areas. There are numerous slender, wavy brown and white cross-lines, and the central region of the wing is often noticeably darker than the rest. The hindwing is similarly coloured, although less strongly marked.
Flight: May–September, in one or two broods.
Caterpillar and life cycle: The caterpillar, about 25mm, ranges from green or grey to rust-coloured, with yellowish intersegmental bands. A narrow dark line runs along the back and there are two yellowish ones on each side, the lower one being quite broad. A typical looper, the caterpillar feeds on various grasses and other low-growing plants, including docks, chickweeds, and dandelions, and often adopts a question-mark shape at rest. Winter is passed as a small caterpillar.
Habitat and range: Almost anywhere. All Europe except the far north.

2 Ghost Swift

Hepialus humuli

Description: Forewing 20–25mm. The upperside is usually pure white in the male: female has yellowish forewings, with reddish brown streaks, and dull brown hindwings. The undersides are dull brown in both sexes.
Flight: June–August, usually at dusk. The male hovers up-and-down and the ghost-like flashing of his white uppersides attracts the female.
Caterpillar and life cycle: The subterranean caterpillar, up to 40mm, is white and fleshy with brown spots and a rust-brown head. It feeds on roots and does considerable damage to crops, including young trees. It sometimes burrows into bulbs and rhizomes. Winter is passed in the caterpillar stage.
Habitat and range: Arable land, parks, gardens, and rough grassland including roadside verges. North and central Europe.

3 Common Swift

Hepialus lupulinus

Description: Forewing 10–12mm: ground colour light to mid-brown, sometimes heavily dusted with white. There are usually two white streaks forming a broad V, although this is often very faint, especially in the female. The antennae are very short, as in all swift moths.
Flight: May–August, usually around dusk: commonly attracted to lights.
Caterpillar and life cycle: The caterpillar, about 35mm, is white or pale yellow with a brown head and faint yellowish spots. It lives in the soil and feeds on the roots of a wide range of grasses and other herbaceous plants, sometimes causing serious damage to field and garden crops and also to lawns. Winter is passed in the caterpillar stage.
Habitat and range: Arable land, parks, gardens, and grassland of all kinds. Most of Europe apart from the south-west.
Similar species: Gold Swift has parallel white streaks instead of a 'V'. Orange Swift has a much more orange male and a much larger female.

1

2/3a

3

1 Leopard Moth

Zeuzera pyrina

Description: Forewing 20–35mm, much larger in female than in male: white and semi-transparent, especially near the tip, with numerous bluish black spots. The hairy thorax has three conspicuous black spots on each side. The male antennae are feathery at the base.
Flight: June–August, often attracted to lights.
Caterpillar and life cycle: The caterpillar, up to 55mm, is cream-coloured with dark spots and a black or dark brown head. It feeds inside the trunks and branches of various deciduous trees and shrubs, sometimes damaging garden and orchard trees.
Habitat and range: Almost anywhere with trees, including town parks and gardens. Southern and central Europe.
Similar species: Puss Moth (p.102) has zig-zag lines instead of spots.

2 Currant Clearwing

Synanthedon tipuliformis

Description: Forewing about 8mm, largely transparent apart from a brownish tip and a large black spot near the middle. Male has four yellow bands on abdomen but female has only three.
Flight: May–August. Usually rests with wings partly open.
Caterpillar and life cycle: The caterpillar, up to 18mm long, is dingy white with shiny yellow spots and a rust-brown head. It feeds inside the stems of currant and gooseberry bushes, causing the leaves to wilt and often bringing about the death of the stems. Winter is passed as a dormant caterpillar. Some individuals may take two years to mature.
Habitat and range: Woods, moors, and gardens, wherever the food-plants grow. Throughout Europe.
Similar species: Red-tipped Clearwing has much redder wing-tips and feeds on sallow.

3 Red-belted Clearwing

Conopia myopaeformis

Description: Forewing 7–10mm, largely transparent except for a large blue-black spot near the middle and a similarly-coloured wing-tip. There is a prominent red band on the abdomen.
Flight: May–August. Usually rests with wings partly open.
Caterpillar and life cycle: The caterpillar, up to 20mm, is creamy white with yellow hairs and a rust-brown head. It feeds inside the trunks and branches of various rosaceous trees, especially apples and pears. Winter is passed in the caterpillar stage and some individuals may take two years to mature.
Habitat and range: Orchards, gardens, and woodland margins. Southern and central Europe: absent from Scotland and Ireland.
Similar species: Large Red-belted Clearwing is a little larger and has a red flush at the base of the forewing.

1

2

3

1 Codlin Moth

Cydia pomonella

Description: Forewing about 10mm: dark grey with irregular lighter cross-bands and often heavily tinged with brown. The outer part of the wing bears a rounded, velvety brown patch that is usually ringed with gold and often encloses small golden streaks as well.
Flight: May–October, in one or two broods.
Caterpillar and life cycle: The caterpillar, up to 10mm, is cream at first, but becomes deep pink later. A major pest of apples, it bores into the fruit and eats both flesh and developing seeds. It also attacks pears and some other fruits. When mature, it leaves the fruit and spins a cocoon under loose bark, where it stays until it pupates in spring.
Habitat and range: Orchards, gardens, parks, and wherever else apple trees grow. All Europe except the far north.

2 Small Magpie

Eurrhypara hortulata

Description: Forewing about 15mm: silky white with greyish brown spots and borders. The thorax is bright yellow with black spots.
Flight: June–August.
Caterpillar and life cycle: The caterpillar, up to 15mm, is flesh-coloured with a brown head and a dirty green line along the back. It feeds on stinging nettle and also on mint and other labiates, rolling or spinning leaves together to form a shelter. Winter is passed as a mature caterpillar in a cocoon spun amongst debris.
Habitat and range: Hedgerows, waste ground, and wherever else stinging nettles grow. All Europe except the far north.

3 Mother-of-Pearl

Pleuroptya ruralis

Description: Forewing about 15mm: pearly white, with grey borders and cross-lines. The hindwing is similar.
Flight: June–August, with a slow, ghost-like flight at dusk.
Caterpillar and life cycle: The pale green caterpillar, up to 12mm, feeds in a rolled-up stinging nettle leaf, where it also spends the winter.
Habitat and range: Hedgerows, waste ground, and wherever else stinging nettles grow. All Europe except the far north.

4 White Plume Moth

Pterophorus pentadactyla

Description: Forewing 12–15mm and split into two feathery lobes. The hindwing is split into three lobes.
Flight: May–August, often coming to lighted windows.
Caterpillar and life cycle: The caterpillar, about 12mm, is bright green with tufts of silvery hairs. It feeds on bindweeds, eating both leaves and flowers. Winter is passed as a dormant small caterpillar.
Habitat and range: Hedgerows, gardens, and waste ground. All Europe.

1/2

3

4

True Flies

Ranging from minute midges to gangling crane-flies and plump bluebottles, the true flies have only one pair of wings. Adult flies feed on liquids, including nectar and blood, but many do not feed at all. Their larvae are all legless grubs or maggots, feeding on a wide range of living and dead materials.

1 Crane-fly

Tipula paludosa

Description: About 25mm long, this very common long-legged fly is also known as the daddy-long-legs. The male has a swollen tip to his abdomen, but the female has a pointed tip which she uses to push her eggs into the soil. The larvae (1a) are the infamous root-feeding leatherjackets. They also come to the surface at night and chew through the bases of stems. Large numbers may live under the lawn.

Food and habits: Adult flies rarely feed. They are on the wing for much of the year, but most common in autumn when they swarm over our lawns and come to lighted windows at night. They rest with their wings outstretched and are commonly seen on house and garden walls.

Habitat and range: Parks, gardens, and grassland of all kinds. All Europe.

Similar species: T. oleracea is very similar but most common in spring.

2 Spotted Crane-fly

Nephrotoma appendiculata

Description: 15–25mm long, with clear, shiny wings. The abdomen is largely yellow with variable black spots.

Food and habits: The adult rarely feeds, but its larva is one of the leatherjackets that do so much damage to roots. Unlike the larger crane-flies, the adult rests with its wings laid flat over its body. It is more often seen resting on leaves than on walls. It flies May–August.

Habitat and range: A wide range of habitats, but particularly common in parks, gardens, and other cultivated places. Most of Europe.

3 Winter Gnat

Trichocera annulata

Description: About 8mm long, including the folded wings, this delicate fly has relatively long legs and a clearly banded abdomen, although this is usually concealed by the wings at rest.

Food and habits: The larvae live in fungi and decaying vegetation and have been known to damage stored potatoes. Adults do not feed. They fly throughout the year, but are most conspicuous on winter afternoons, when the males form large dancing swarms over garden ponds, parked cars, and other prominent objects. The swarms usually develop in weak afternoon sunshine, and can be seen even when there is snow on the ground.

Habitat and range: Almost everywhere. All Europe.

Similar species: There are several very similar species, but the abdomen is usually unbanded.

1

1a

2/3

1

Bibio hortulanus

Description: 8–12mm, including the folded wings. The latter are smoky brown at the front but otherwise very clear. Female is chestnut brown or brick-coloured above, while the smaller male is black and has much larger eyes. The antennae are short and stout in both sexes.
Food and habits: The nectar-feeding adults are important pollinators of apples and other fruit trees. They fly in spring and mating pairs often bask on house walls. The grubs feed mainly in leaf litter and other rotting material, but may cause minor damage to crop roots.
Habitat and range: Open woods, grassland, and cultivated areas. Much of Europe except the far north.

2

Bee-fly

Bombylius major
Description: 10–12mm, with a long beak or proboscis, slender legs, and a furry brown coat. The front edge of each wing is dark brown.
Food and habits: Feeds on nectar, which it takes from the flowers with its beak: often hovers with a high-pitched whine in shafts of sunshine. One the wing March–June, it is quite harmless. The larvae live as parasites in the nests of solitary bees and wasps (see p.148 and p.154).
Habitat and range: Light woodland and most open habitats. All Europe but rare in the far north.

3

Hover-fly

Syrphus ribesii
Description: About 10mm long, this is one of many harmless, yellow and black hover-flies that mimic wasps. As in all hover-flies, the outer veins turn and run parallel to the outer edge of the wing, forming what is known as a false margin.
Food and habits: Adults are on the wing March–November. They feed mainly on nectar and honeydew, but may also crush and swallow pollen grains. Although legless, the slug-like green larvae are active predators of aphids and valuable allies in the garden.
Habitat and range: Flower-rich places of all kinds: very common in parks and gardens. Throughout Europe.
Similar species: S. vitripennis is almost identical and equally common. Many others have similar colours, but slightly different patterns.

4

Hover-fly

Episyrphus balteatus
Description: 10mm. The first two broad black bands on the abdomen are each usually followed by a much narrower black band.
Food and habits: Adults eat nectar, pollen, and honeydew. On the wing for most of the year, often migrating in huge swarms. The larvae eat aphids.
Habitat and range: Anywhere with flowers: abundant in gardens. All Europe.

1

2/3

4

1 Hover-fly

Scaeva pyrastri

Description: 12–15mm. The white or cream crescent-shaped marks on the abdomen are narrower in the middle than at the ends and the inner arm of each crescent extends further forward than the outer one.
Food and habits: Adults feed mainly on nectar and honeydew and fly May–November, although most abundant in late summer. The larvae eat aphids.
Habitat and range: Flower-rich places of all kinds: very common in parks and gardens. Most of Europe, but rare in Scotland and the far north.
Similar species: S. selenitica has comma-shaped abdominal marks tapering strongly towards the outside.

2 Drone-fly

Eristalis tenax

Description: 10–15mm long, this hover-fly is named for its resemblance to a honey bee drone (see p.156). A wide, dark stripe runs down the middle of the face and the hind legs bear brush-like hairs. The pale abdominal markings vary and are sometimes missing.
Food and habits: Feeds mainly on nectar and pollen. On the wing all the year and, although dormant in the coldest weather – often in houses – it often basks on sunny walls in the middle of winter. The larva lives in muddy ditches and other stagnant water. It has a telescopic breathing tube and is called a rat-tailed maggot.
Habitat and range: Flower-rich places of all kinds: often abundant in parks and gardens. Throughout Europe.
Similar species: E. horticola (below) and several other hover-flies have similar patterns, but no hair-brushes on the hind legs.

3 Hover-fly

Eristalis horticola

Description: 10–12mm, with variable bright yellow abdominal patches and a small dark patch on each wing. The face has a narrow black stripe.
Food and habits: Feeds on pollen and nectar and flies April–October. The larva is like that of the Drone-fly (above).
Habitat and range: Flowery places, especially around shrubs and trees. Much of Europe except the far north.
Similar species: Drone-fly (above) is darker and has hairy back legs.

1

2

3

1 Narcissus-fly

Merodon equestris

Description: 10–15mm, with black legs and a deep U-shaped bend in the third long vein of the wing. The plump body is clothed with grey, brown, or black hairs arranged in several different patterns. The various forms of this hover-fly mimic various kinds of bumble bee.

Food and habits: Adults feed on nectar and pollen and are often seen on dandelion flowers. They fly March–August, but are most common in May, when the females bask on the withering leaves of daffodils and other bulbous plants. Eggs are laid on the leaf bases and the resulting grubs tunnel into the bulb. Even if the bulbs are not completely destroyed, their flowering capacity is greatly reduced.

Habitat and range: Woods, hedgerows, parks, and gardens. All Europe except the far north.

Similar species: Eristalis intricarius has similar patterns but its legs are partly pale. Some species of *Volucella* are superficially similar but have no U-shaped bend in the veins. Note: the grubs of several other flies can be found in bulbs.

2 Hover-fly

Volucella zonaria

Description: 15–25mm, with chestnut patches at the front of the abdomen and an extensive yellowish brown tinge on the basal half of the wing. The fly bears an uncanny resemblance to a hornet (see p.152), especially in flight, when it emits a hornet-like buzz.

Food and habits: Adults feed on pollen and nectar and also on honeydew. On the wing May–October, they are most noticeable in late summer, when they often enter houses. The larvae live as scavengers in wasps' nests.

Habitat and range: Woodland margins, hedgerows, waste ground, parks, and gardens: often in urban areas. Southern and central Europe, including southern England.

Similar species: V. inanis is smaller, with yellow patches at the front of the abdomen.

3 Cluster-fly

Pollenia rudis

Description: 8–10mm. Golden or silvery hairs cover the thorax, although they may fall with age. The abdomen appears largely black from some angles, but from others it reveals a chequered black and grey pattern.

Food and habits: Adults probably feed mainly on nectar and other sweet liquids, but the larvae are internal parasites of earthworms. Adults can be found at all times of year but are most obvious in autumn, when they often gather in enormous numbers on sunny walls and then make their way into sheds, lofts, attics, and other sheltered places for hibernation.

Habitat and range: Most habitats except heaths and mountains, where the worm population is low: very common on grassland and all cultivated areas. Throughout Europe.

1

2

3

1 Flesh-fly

Sarcophaga carnaria

Description: 12–20mm, with red eyes and very large feet. The abdomen has a chequered black and grey pattern that varies with the angle from which it is viewed.
Food and habits: Adults feed on nectar, rotting carrion, and dung. On the wing throughout the year, they often bask on sunny walls in winter. Females give birth to active grubs that feed in dung and carrion.
Habitat and range: A wide range of habitats: common around houses but rarely found indoors. All Europe.
Similar species: There are several very similar *Sarcophaga* species.

2 Greenbottle

Lucilia caesar

Description: 8–15mm, with a metallic green body and silvery jowls below the eyes. The fourth long vein is sharply bent.
Food and habits: Adults feed on nectar and carrion juices. They fly at all times of the year, including mid-winter, when they often bask on sunny walls. The maggots feed on carrion, including scraps of meat and fish consigned to the dustbin.
Habitat and range: Most habitats: common around houses, but rarely seen indoors. All Europe.
Similar species: Several other flies have metallic green bodies, but some have green or yellow jowls.

3 Cabbage Root-fly

Delia radicum

Description: 5–7mm: a bristly black or dark grey fly in which the outer part of the third long vein curves gently backwards. The fly is most often seen when it is dead. It is highly susceptible to attack by a fungus, and clusters of dead flies are frequently seen on cabbages and other plants, bound to the leaves by strands of the murderous fungus. Such sights are most common in late summer and autumn.
Food and habits: Adults fly March–November and feed mainly on nectar. The grubs feed on the roots of cabbages and other brassicas, causing the leaves to become limp and yellow. Whole plants are quickly killed by heavy infestations.
Habitat and range: Wherever cruciferous plants grow, but most abundant on cultivated land. The species seems to have increased recently, possibly as a result of the increased cultivation of rape, which is a common larval food-plant. All Europe.
Similar species: There are many similar flies, including the Onion-fly whose maggots tunnel into onions, causing the leaves to wilt and often causing complete bulbs to rot.

1

2

3

Sawflies

Sawflies belong to the same major group as the bees and wasps, but they have no 'wasp-waist'. They are named for the saw-like ovipositors of many species, which the females use to cut slits in plants before laying their eggs in them. Adult sawflies feed mainly on pollen and most of them are rather shy insects. The gardener is more likely to see their larvae. Most of these feed on leaves and several species cause appreciable damage in the garden. Some of them can be mistaken for the butterfly or moth caterpillars, although they have at least 6 pairs of fleshy abdominal legs: true caterpillars have no more than 5 pairs.

1 Gooseberry Sawfly

Nematus ribesii

Description: Up to 10mm, the adult is largely yellow, although the male abdomen has a variable amount of black. But it is the spotty, greyish green larva, up to 35mm, that is most often seen.

Food and habits: The larvae feed gregariously on the leaves of gooseberries and currants in the summer and can strip the bushes bare. Pupation takes place in the soil and the adults fly April–September.

Habitat and range: Wherever the food-plants grow, but most common in gardens. Most of Europe except the far north.

Similar species: Pristiphora pallipes also attacks currants and gooseberries, while the very similar *P. alnivora* defoliates aquilegias. There are many Similar species, each with its preferred food-plant.

2 Solomon's Seal Sawfly

Phymatocera aterrima

Description: Up to 9mm, with a black body and smoky black wings.

Food and habits: Adults fly May–June and feed at various flowers, but are most often seen crawling on or flying slowly over Solomon's Seal plants. The larvae, up to 20mm, are grey with black heads. They feed gregariously and quickly reduce the leaves to bare veins. Pupation takes place in the soil.

Habitat and range: Woods and gardens, wherever Solomon's Seal grows. Southern and central Europe.

3 Cherry Slug Sawfly

Caliroa cerasi

Description: About 5mm long, the adult is completely black, but it is rarely noticed. It is the slug-like black larvae, up to 15mm, that usually come to the gardener's attention.

Food and habits: Also known as the Pear-Cherry Slugworm, the larva feeds on the leaves of pears, cherries, apples, and many other rosaceous trees. It rasps away the upper layers of a leaf, leaving just the veins and the lower surface – a process sometimes called skeletonisation. The larvae are active May–September. Pupation takes place in the soil.

Habitat and range: Woods, parks, and gardens, wherever there are suitable food-plants. Southern and central Europe.

1

2/2a

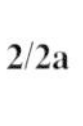

3

1 Rose Slug Sawfly

Endelomyia aethiops

Description: About 5mm, the adult is completely black apart from the pale patches on its front and middle legs. It is rarely noticed, but the damage done by its larvae is only too often seen on garden roses. The slender, slug-like larva is up to 15mm and is yellowish green with a brown head.
Food and habits: The larvae feed gregariously on the leaves of both wild and cultivated roses, attacking either the upper or lower surface. They remove the surface layers and leave only the veins and the skin on the opposite side, so the leaf looks as if it has small windows in it. They are active May–September and they pupate in the soil.
Habitat and range: Hedgerows, parks, gardens, and anywhere else that roses grow. Throughout Europe.

2 Birch Sawfly

Croesus septentrionalis

Description: Measures 8–10mm, with a broad red band on black abdomen. The wings are clear, with a black mark on the front edge of the forewing and dark smudge just behind it. Adults are rarely noticed, but the greyish green larvae are abundant on birches and other trees: up to 25mm and heavily-spotted with black, they have a black head and yellow legs.
Food and habits: Adults feed at flowers May–September and lay their eggs in the leaf veins of the food-plants. The larvae feed gregariously around the edges of their leaves. If disturbed, they raise their abdomens and curve them forward over their heads. This position is typical of many sawfly larvae. The larvae also release a rather pungent odour.
Habitat and range: Mainly on birches and alders, including park and garden trees: also on hazel, sallow and poplar. Throughout Europe.

3 Horntail

Urocerus gigas

Description: Up to 40mm, including the female's stout ovipositor that gives this large sawfly its name. The female, pictured here, is black and yellow, but the male, which is slightly smaller, is black and orange. The insects are also called woodwasps.
Food and habits: The female's ovipositor is more like a drill than a saw and she uses it to drill into pine trunks. The legless grubs feed on the timber for several years and also pupate there.
Habitat and range: Mainly in and around coniferous woodland, but adults sometimes appear on new housing estates. The larvae take several years to grow up and can survive even after the trees have been converted into floorboards and rafters. The adults cause great alarm when they appear in or around the house but, despite their similarity to hornets (see p.152), they are quite harmless.

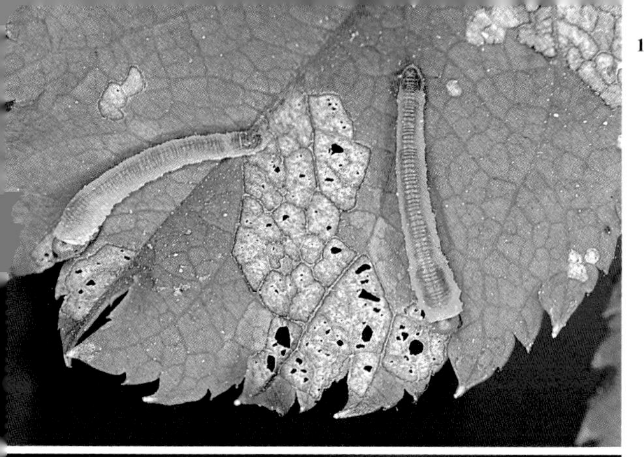

1

2

3

Ichneumons

A large group of parasitic insects whose grubs grow up in or on other insects, especially caterpillars of butterflies and moths. The female parasites lay their eggs either on or in the host bodies, which they usually find by scent. There is commonly just one ichneumon grub to each host insect, but some small species may have over 100 grubs inside a single host. The host insects remain alive until the ichneumon grubs reach full size, but then they die and the ichneumons pupate inside or outside the shrivelled skins. Adults vary greatly in appearance but all have relatively long antennae.

1 Yellow Ophion

Ophion luteus

Description: 15–20mm, rusty brown with a strongly arched abdomen markedly flattened from side to side. The female's short ovipositor can pierce the skin and cause a good deal of pain if the insect is handled.
Food and habits: Adults take nectar and pollen from umbellifers and other flowers July–October, and are attracted to lighted windows. Larvae are internal parasites of various caterpillars, usually just one parasite per host.
Habitat and range: Almost anywhere well-vegetated habitat. Most of Europe except the far north.
Similar species: Many similar ichneumons; some have dark tip to abdomen.

2 *Amblyteles armatorius*

Description: About 15mm, this ichneumon is largely black but has a prominent yellow triangular scutellum at the rear of the thorax and yellow bands on the abdomen.
Food and habits: Adults feed on pollen and nectar, especially from umbellifers. They are on the wing June–October, and then go into hibernation until the spring. The larvae grow up in the caterpillars and pupae of various large moths, with just one parasite per host. They always emerge from the host pupa.
Habitat and range: A wide range of habitats, including woods and moorlands as well as gardens. Throughout Europe.
Similar species: There are several very similar ichneumons.

3 *Apanteles glomeratus*

Description: 4mm, largely black with smoky brown wings. The adults are rarely noticed, but their larvae play a major role in controlling Large White and Black-veined White butterflies.
Food and habits: Adults are on the wing May–September in two broods. The female lays numerous eggs in a young caterpillar, where the grubs feed until they are fully grown – at which time they emerge and spin small golden cocoons around the shrivelled skin of the host. A new generation of *Apanteles* adults emerges from the cocoons in due course.
Habitat and range: Many habitats, wherever the host species occur, but most common in cultivated areas. Throughout Europe.

1

2

3

1 Ruby-tailed Wasp

Chrysis ignita

Description: 7–10mm, with a brilliant red or purple abdomen. The head and thorax are blue or green, often with a golden sheen.
Food and habits: Adults take nectar from various flowers April–September. Larvae live in the nests of mason wasps (see p.148) where they feed on the wasp grubs and also stored food. Adults often seen running over walls and tree trunks looking for suitable wasps' nests in which to lay their eggs.
Habitat and range: A wide range of open habitats. Throughout Europe.
Similar species: Several similar species, often known as cuckoo wasps due to their breeding habits. Some use solitary bees as hosts instead of wasps.

Ants

The ants are all social insects, living and working together in colonies. Each colony contains one or more egg-laying queens and hundreds or even thousands of wingless workers. The latter collect food, rear the grubs, and do all the other work. They are the only ants normally seen outside the nest except during the mating flights. Most European ants are omnivorous, feeding on nectar, aphid honeydew (see p.66), seeds, and other living or dead insects. Spectacular mating flights occur in the summer – usually on still, humid days – when the 'flying ants' pour out from the nests. Larger than the workers, these are the new queens and males and they fly off like plumes of smoke. After mating males die and females that are not eaten by birds break off their wings and either start new colonies or go into existing ones. Active for most of the year, becoming dormant in their nests in winter.

2 Black Garden Ant

Lasius niger

Description: Up to 7mm (worker): black or dark brown with a waist consisting of a single scale-like segment. This ant has no sting.
Food and habits: Omnivorous, with a strong liking for honeydew and other sweet things. It often 'milks' aphids on plants and may make tubular collars of soil around the stems of roses and other plants to protect the aphids. It even installs them on roots in its nests. The latter are often under stones and paths – even town pavements – and sometimes in or under house walls. One queen and several thousand workers live in a colony. Mating flights occur in July or August.
Habitat and range: Most fairly open habitats. Throughout Europe.

3 Red Ant

Myrmica rubra

Description: Up to 7mm (worker): chestnut brown with a 2-segmented waist. This ant, pictured here with its grubs, can sting.
Food and habits: Omnivorous, but less interested in honeydew than the black ant and with a bias towards animal food. It nests under stones or in rotting logs and its colonies contain one or more queens and only a few hundred workers. Mating flights occur in July or August.
Habitat and range: A wide range of open habitats. Throughout Europe.
Similar species: *M. ruginodis*, equally common, is almost identical.

1

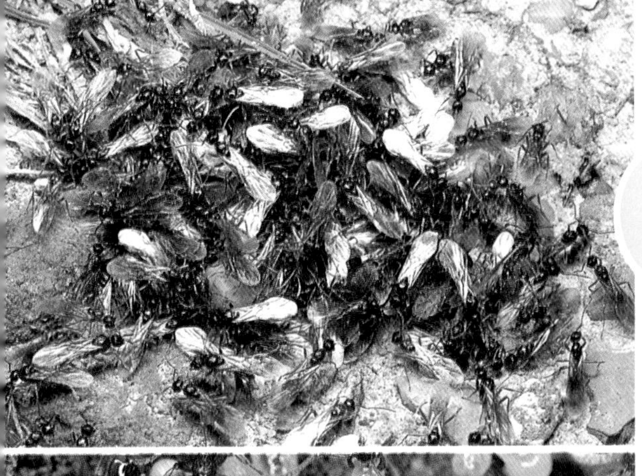

2

3

1

Digger Wasp

Ectemnius cavifrons

Description: 10–16mm: one of several very similar solitary wasps with black and yellow abdomens and very square heads. The species are not easy to separate. Unlike the mason wasps (below) and social wasps (p.150–53), they all lay their wings flat over the body at rest.
Food and habits: Adults visit flowers, especially umbellifers, June–October. They feed on pollen and also collect flies. The latter, paralysed by the wasp's sting, are packed into burrows in old fence posts and other soft timber, as pictured here, and serve as a living larder for the wasp's grubs. Small hover-flies, such as *Episyrphus balteatus* (p.132), are the main prey.
Habitat and range: A wide range of habitats: one of the commonest of the digger wasps, often found on woodland margins and in hedgerows and garden log piles. Southern and central Europe, but not Scotland.

2

Mason Wasp

Ancistrocerus parietinus

Description: 10–14mm: one of several rather similar mason wasps with a more or less square black mark on the yellow band at the front of the abdomen. The latter is egg-shaped and smoothly rounded at the front. As in the social wasps (see p.150–53), the wings are folded lengthways along the sides of the body at rest, with the top of the abdomen exposed.
Food and habits: Adults feed mainly on nectar and honeydew April–August. They construct nests with mud or clay and tuck them into all sorts of cavities, including holes excavated in loose mortar and then 're-pointed' when the nests are complete. The nests are stocked with small caterpillars that have been paralysed by the wasp's sting.
Habitat and range: Almost anywhere, but most common in areas with rocks or walls in which the nest can be built. Most of Europe.
Similar species: A. parietum and *A. nigricornis*, with similar patterns and habits, can be distinguished only by detailed examination.

3

Potter Wasp

Delta unguiculata

Description: 15–20mm, brown and yellow with a narrow, bell-shaped waist sharply separated from the pear-shaped abdomen.
Food and habits: Adults feed on nectar and honeydew June–August. They make nests with fine sand or clay, and can often be seen gathering it from damp ground, as pictured here. The material is moistened with saliva, stuck to rocks or walls, and fashioned into hollow, pea-sized cells. Each cell receives an egg and a number of small caterpillars before it is sealed up. Several cells are often built together. When dry, they are easily mistaken for simple lumps of mud.
Habitat and range: A wide range of habitats, including gardens: most often seen near water. Southern and central Europe, but not the British Isles.

1

2

3

Social Wasps

Live in annual colonies, each started by a queen who lays all the eggs. The colony inhabits a nest built with paper made by chewing up wood fragments. A mature nest may contain several thousand 6-sided cells in which the grubs are reared. Nearly all wasps in the nest are workers, smaller than the queen, but otherwise very similar. They collect food and building materials and do all the work in the nest. Adults eat nectar, fruit, and other sweet things, but the grubs are reared on animal food – especially other insects. The inhabitants of a single nest consume thousands of insect pests in a summer, so wasps are really very useful. Workers are on the wing from late spring until the first frosts. Males, identified by long antennae, occur only in late summer, when they mate with the new queens. Colonies break up in the autumn and only newly-mated queens survive the winter, hibernating in sheltered places, emerging in spring. Apart from the hornet (p.152), all European species are black and yellow. They fold their wings lengthways along the sides of the body at rest. Apart from males, all wasps can sting, but do not usually attack unless annoyed. These descriptions refer to workers and queens: males often have slightly different patterns.

1 Common Wasp

Vespula vulgaris

Description: 9–18mm. Parallel-sided yellow streak each side of the thorax, four yellow spots at rear. Usually has a black anchor-shaped mark on face.
Nesting habits: Usually nests in hedgebanks and other well-drained underground sites, but often in wall cavities and roof spaces. The nest is roughly spherical and covered with shell-like lobes of paper. Made from fairly rotten wood, the paper tends to be yellow and quite brittle.
Habitat and range: Common in most habitats. Throughout Europe.
Similar species: See below.

2 German Wasp

Vespula germanica

Description: Very similar to Common Wasp (above), but thoracic stripes bulge in the middle and face usually has three black dots.
Nesting habits: Nests in similar places to Common Wasp but nest paper, made from sounder wood, is greyer and less brittle.
Habitat and range: Common in most habitats. All Europe except far north.

3 Norwegian Wasp

Dolichovespula norvegica

Description: 11–17mm: like Common Wasp but thorax has only 2 yellow spots at the rear. Face has a broad black vertical bar.
Nesting habits: Nests are smooth and grey and are hung in bushes. Fairly small, they house no more than a few hundred wasps.
Habitat and range: Anywhere with trees and shrubs. Most of Europe; most common in upland and northern areas. Scarce in east and central England.
Similar species: Tree Wasp, which has a similar nest, has no more than a small black dot in the middle of its face.

1

2

3

1 French Wasp

Dolichovespula media

Description: 15–22mm: resembles Common Wasp, but often has more black on abdomen and sometimes has only 2 yellow spots on rear of thorax. Face has a slim black vertical bar in the middle.
Nesting habits: The greyish, ball-shaped nest has a fairly smooth surface and is usually hung in trees and bushes. Young nests have a 'spout' at the bottom.
Habitat and range: Woods and other areas with trees and bushes. Most of Europe except far north: unknown in the British Isles until 1980, but now spreading rapidly and not uncommon in town gardens.

2 Hornet

Vespa crabro

Description: 20–30mm, with a chestnut brown thorax and a brown and yellow abdomen. Queens are much larger than any other wasps.
Nesting habits: Nests are usually built in some kind of cavity, including hollow trees, wall cavities, and disused chimneys. The paper, made from rotten wood, is usually yellowish or pale brown. A mature colony contains several hundred workers. Adults are fond of sap oozing from damaged tree trunks, and sometimes harm trees by gnawing through the bark of young shoots to reach the sap. They rear their young on a wide range of insects, including butterflies and dragonflies. The nests of bumble bees and honey bees are also systematically plundered for prey. Wings are usually discarded before the prey is taken back to the hornet nest. The insects often fly at night in warm weather.
Habitat and range: Mostly in wooded areas, including parkland but regular visitors to gardens. Most of Europe except the far north, but absent from Scotland and Ireland. Populations fluctuate a great deal in the British Isles.
Similar species: Some large hover-flies (p.136) mimic the hornet, but all have short antennae and only one pair of wings.

3 Paper Wasp

Polistes gallicus

Description: 10–15mm: superficially similar to the other social wasps, but the abdomen is hairless and it tapers smoothly towards the front instead of being squared off.
Nesting habits: The nest is made of greyish paper and consists of a single layer of cells that are freely exposed to the air. The nest is usually attached to vegetation or to rocks. There are rarely more than a few dozen cells, and rarely more than a handful of adults at any one time. The grubs are fed mainly on chewed caterpillars.
Habitat and range: Most habitats, but uncommon in dense woodland and in very open areas: frequently nests on buildings, especially in porches and under other over-hanging structures. Southern and central Europe: most common in the south. A rare vagrant to the British Isles.
Similar species: There are several very similar species, all with varying abdominal patterns and difficult to distinguish.

1

2

3

Bees

Bees include both solitary and social species, the latter living in colonies of queens and workers and all working together for the good of the colony – just like the social wasps (p.150–53). But the bees build their nests with wax from their own bodies. Adult bees feed on nectar and pollen and, unlike the wasps, they also collect these foods for their grubs. Generally hairier than wasps, their hairs help them to collect pollen. Bumble bees and worker honey bees are often seen with blobs of pollen on their back legs held in place by fringes of stiff hairs known as pollen baskets. Honey bees sting in defence of their colonies, but other bees rarely sting unless handled.

1 Tawny Mining Bee or Lawn Bee

Andrena fulva

Description: 10–12mm. The female is easily recognised by her bright chestnut abdomen. Male is dark brown or black and rarely noticed.

Food and nesting habits: The female is very fond of flowers of currants and gooseberries. On the wing April–June, this solitary species nests in the ground, especially on bare areas of lawns, and throws up a little mound of soil like a miniature volcano around the mouth of its burrow. The nest, like that of other bees, is stocked with pollen and nectar.

Habitat and range: Most open habitats, including parks and gardens, light woodland, and scrub. Southern and Central Europe, including southern Britain, but absent from much of the south-west.

2 Mason Bee

Osmia rufa

Description: 8–12mm long and rather plump, with a black head and reddish brown hair on the abdomen. The female, pictured here, has denser hair than the male and has a pair of short, curved horns just below her antennae. Male is much smaller than the female but has longer antennae.

Food and nesting habits: Feeds at a wide range of flowers April–July. The nest, consisting of several mud cells, is wedged into all kinds of holes and crevices. Walls with soft mortar sometimes attract large numbers of these bees although they rarely do any serious damage.

Habitat and range: Almost anywhere with flowers and suitable nest holes: very common in parks and gardens. Southern and central Europe.

3 Leafcutter Bee

Megachile centuncularis

Description: About 10mm: dark brown above with a conspicuous hollow at the front of the abdomen. Female has a prominent patch of orange hair, used for holding pollen, on the underside of the abdomen.

Food and nesting habits: Adults visit a wide range of flowers May–August. The female uses her jaws to cut round or oval sections from the leaves and carries them back to her nest cavity, usually in a deep crevice, where she uses them to make sausage-shaped cells for her grubs. She is particularly attracted to roses and may use petals as well as leaves.

Habitat and range: Woods, hedgerows, parks, and gardens. All Europe.

1

2

3

1 Flower Bee

Anthophora plumipes

Description: 12–15mm. Female is black with orange pollen brushes on hind legs: male, seen here, is largely brown with fans of long hairs on middle legs.
Food and nesting habits: Adults regularly hover in front of lungwort and other tubular flowers March–June, sucking nectar with the long tongue. They resemble bumble bees (p.158), but flight is much faster and flight tone has a much higher pitch. The nest is a cluster of small clay cells and is normally built in steep banks, although some bees nest in old walls, excavating the soft mortar just like the Mason Bee (p.154).
Habitat and range: Well-drained habitats of all kinds, but most abundant in and around human settlements. Most of Europe, but not Scotland.

2 Violet Carpenter Bee

Xylocopa violacea

Description: 20–25mm, with a shiny black body and very hairy legs. The dark wings display rich blue and violet reflections at certain angles.
Food and nesting habits: Adults feed from numerous flowers in late summer and autumn, and again in spring after hibernation in holes in walls and timber. Mating takes place in the spring and the females excavate nest burrows in dead wood, including fence posts and building timbers. The cells of the nest are usually arranged vertically, one above the other and separated by layers of sawdust. The new bees emerge in the summer. Despite their large size and fast, noisy flight, the bees are not aggressive and rarely sting.
Habitat and range: Woodland margins, gardens, and wherever else there is dead wood for nesting. Southern and central Europe, occasionally wandering to the British Isles and other more northerly areas.
Similar species: Three more very similar species live in western Europe.

3 Honey Bee

Apis mellifera

Description: 12–15mm (queens about 20mm but never seen foraging). The abdomen is quite plump and mainly brown, but there are numerous races of honey bees and many have extensive orange areas at the front of the abdomen. The surest way to separate honey bees from other bees is to look for the very long narrow cell near the wing-tip.
Food and nesting habits: Pollen and nectar are gathered from a wide range of flowers from early spring to late autumn. A social bee, it lives in permanent colonies with a single queen and several thousand workers. Small numbers of males (drones) occur in spring and summer. They are stouter than workers and have longer antennae. Most honey bees live in artificial hives, but wild colonies exist in hollow trees and other sheltered sites. The nest contains several vertical wax combs, covered with hexagonal cells used for rearing grubs and also for storing pollen and honey. Queen and workers are dormant in the winter.
Habitat and range: Almost anywhere. A native of southern Asia, but cultivated strains are now almost worldwide.

2

3

Bumble Bees

Bumble bees are fairly large, hairy social bees. They form annual colonies like those of the wasps (pp.150–53). The colonies start to break up in late summer and only young, mated queens survive to start new colonies in the spring. The nests are usually under the ground, often in old mouse holes, although some species nest in thick grass at ground level. The nest is a ball of grass or moss, inside which the queen makes a few wax cells. She rears the first batch of workers herself and, because she can collect only a limited amount of food, the early workers are very small. Later workers grow larger because there are older ones to collect food for them. Queens are common in our gardens in early spring, but most of the bumble bees seen at other times are workers. Bumble bee nests rarely contain more than a few hundred workers and some colonies are much smaller. Bumble bees are mostly quite docile insects and sting only if molested. Queens and workers usually have similar patterns but males, which are much less numerous and less often seen than workers, may have different colours and patterns.

1 Red-tailed Bumble Bee

Bombus lapidarius

Description: Up to 25mm: completely black apart from the red tip of the abdomen. Pollen baskets on hind legs are black.
Food and nesting habits: On the wing from April, it favours clovers and purple-flowered composites. Queens are common on blackthorn in the spring. Nests on or under the ground, very often under large stones.
Habitat and range: Occupies many habitats, but most common in open areas. Most of Europe, but often confined to coastal areas in the north.
Similar species: Psithyrus rupestris, its cuckoo bee parasite (see p.160), is much shinier. *B. ruderarius* has brown pollen baskets.

2 White-tailed Bumble Bee

Bombus lucorum

Description: Up to 25mm. The collar (front of thorax) and 2nd abdominal segment are yellow and the tip of the abdomen is white. Female seen here.
Food and nesting habits: On the wing as early as February, the queens are very fond of sallow catkins. Workers visit a wide range of flowers. Nests below ground, often making use of the shelter afforded by garden sheds.
Habitat and range: Abundant in flowery places throughout most of Europe.

3 *Bombus hortorum*

Description: Up to 24mm. Collar, rear edge of thorax, and 1st abdominal segment are yellow: tip of abdomen is white. The hair is longer than in other garden species, giving the insect a slightly scruffy appearance.
Food and nesting habits: The bee has a very long tongue and specialises in deep-throated flowers, including many garden flowers. Queens enjoy white deadnettle in spring. Usually nests on or just under the ground.
Habitat and range: Common in gardens throughout Europe.
Similar species: Psithyrus barbutellus, a cuckoo bee (see p.160) that parasitises *B. hortorum*, is less hairy and has a very shiny abdomen.

1

2

3

1 *Bombus pascuorum*

Description: Up to 18mm. Thorax is usually reddish brown (the only garden bumble bee with this colour), although very dark in northern areas. Abdomen has a rather thin coating of brown, black, or grey hairs.
Food and nesting habits: Visits a wide range of wild and cultivated flowers: very fond of lavender. Nests are usually built in long grass at ground level, but this species also builds in old birds' nests well above ground and even takes up residence in bird boxes. Queens usually appear late March or April and the colonies continue well into the autumn, long after most other bumble bees have disappeared.
Habitat and range: Most habitats, but uncommon in the more exposed places: a very common garden species. All Europe.

2 Buff-tailed Bumble Bee

Bombus terrestris

Description: Up to 25mm. Collar and 2nd abdominal segment orange or golden yellow – usually much darker than in White-tailed Bumble Bee (see p.158). Queen has tip of abdomen buff or gingery in the British Isles, but white on the continent. Workers usually have a white tip, often tinged with brown towards the front.
Food and nesting habits: Visits a wide range of flowers, including apple and cherry blossom. Queens are frequent visitors to sallow catkins in March and April. Nests are built well below ground level.
Habitat and range: Almost any well-vegetated habitat. Most of Europe, but absent from northern Scotland and the far north.

3 Cuckoo Bee

Psithyrus vestalis

Description: Up to 22mm. Collar is golden and tip of abdomen is white with patches of yellow at the sides. The rest of the body is black and shiny, with sparse hair. The wings are noticeably brown. A male is pictured here. Females have furry hind legs with no pollen baskets.
Food and nesting habits: Visits numerous flowers but is especially attracted to thistles, knapweeds, and other purple flowers. One of the cuckoo bees, it makes no nest and has no workers. Females lay eggs in the nests of Buff-tailed Bumble Bee and often kill the rightful queens. The worker bumble bees then rear the cuckoo bee grubs and new cuckoos appear in mid-summer. Mated females hibernate and do not wake until the host nests are well established in the spring. Each cuckoo bee species has its own host species and usually resembles it quite closely.
Habitat and range: Almost anywhere inhabited by Buff-tailed Bumble Bee, but rare in the north and probably absent from Scotland and Ireland.
Similar species: Buff-tailed Bumble Bee is less shiny, with a golden band at front of abdomen: wings are clearer.

1

2

3

Spiders

Spiders are commonly confused with insects, but they have eight legs instead of six and their bodies are divided into two sections instead of three. Spiders never have wings. Entirely carnivorous, they consume large numbers of flies and other pests. All spiders produce silk, but not all of them make webs: many are free-roaming hunters.

1

Garden Spider

Araneus diadematus

Description: Male 4–8mm: female 10–15mm. Both sexes have a clear white cross of irregular-sized dots on the abdomen. Ground colour ranges from bright orange to black, but is usually some shade of brown.

Habits: Makes a more or less vertical orb-web, usually well off the ground on bushes and fences, or between the branches of trees. It catches a wide range of flying insects. The spider usually hides close to the web, but sometimes sits in the middle. Adult late summer and autumn, when the webs are most obvious. Overwinters in the egg stage, the eggs being enclosed in silken bags fixed in bark crevices or under window-sills. Swarms of tiny yellow and brown spiderlings emerge from the bags in the spring.

Habitat and range: Almost any well-vegetated habitat: very common in hedgerows and gardens. Throughout Europe.

2

Araniella cucurbitina

Description: 3–6mm. The front half of the body is brown and often quite shiny. The abdomen is pale green, with a yellow stripe on each side and a number of small black dots. A red patch surrounds the spinnerets just under the rear end.

Habits: Spins a very small orb web, often across the surface of a single leaf, and often sits in the middle of it. Feeds on small flies and aphids. Adult mid-summer to autumn.

Habitat and range: Trees, bushes, and tall herbaceous plants, usually more than 1.5m above the ground. Abundant in orchards, shrubberies, and herbaceous borders. Throughout northern and central Europe.

Similar species: A. opisthograpta is indistinguishable without microscopic examination. Other species live in Europe but occur mainly in woodland.

3

Nuctenia umbratica

Description: 8–15mm, with a clearly flattened abdomen. Usually very dark, but the sides of the abdomen may be lighter brown, so that the pale-edged leaf-like pattern (the folium) shows up more clearly.

Habits: Active mainly by night, spinning an orb-web on tree trunks, fences, and walls, especially around doorways – from which the spider hangs menacingly at about head height. Hides in crevices by day; usually spins a fresh orb-web each evening. Feeds mainly on moths and flies. Adult all year.

Habitat and range: Woods, orchards, and gardens – mainly on sheds and fences. Throughout northern and central Europe.

1

2

3

1

Meta segmentata

Description: 4–8mm, the two sexes being more or less equal in this species. The colour varies from pale to dark brown, usually with pale triangles near the front of the abdomen, and there is a fairly clear dark mark resembling a fork or an anchor just behind the head.
Habits: Spins an orb-web, usually sloping noticeably from the vertical, in hedges and other vegetation up to 2m above the ground. The web has no central platform and the spider usually hides under a nearby leaf. The male usually gives the female a 'wedding present' of a silk-wrapped fly before mating with her. Adults most noticeable in late summer and autumn, but some are found in spring – suggesting they may hibernate in some places.
Habitat and range: Abundant in all kinds of low-growing vegetation, including garden borders. Throughout Europe.

2

Zygiella x-notata

Description: 3–7mm. Ground colour ranges from grey to dark brown, with a distinct leaf-like pattern on the abdomen. This usually has a silvery lustre and is often outlined in pink.
Habits: Spins a more or less vertical orb-web, usually on buildings or fences. The corners of doors and window frames are favourite sites and the web is easily recognised by the two empty sectors near the top. The spider usually hides in a nearby crevice and waits for flies and other insects to become trapped in the web. Adult throughout the year.
Habitat and range: Rarely found away from buildings in northern and central Europe: less tied to human habitation in the south. All Europe.
Similar species: Z. atrica has a more silvery abdomen with two chestnut spots near the front, but is rarely found on buildings.

3

Argiope bruennichi

Description: Male 4–5mm: female 10–25mm. Female is easily recognised by the black, yellow, and silver abdominal bands. Male is rarely seen.
Habits: Spins a more or less vertical, wheel-shaped orb-web in long grass or other herbage, often at the base of hedgerows, and usually adds a zig-zag band of silk that runs vertically through the centre – although this band is often absent from the webs of mature spiders. Grasshoppers are the major food in most places. The spider is usually seen sitting head-down at the centre of the web. Adult mid-summer to autumn. Overwinters in the egg stage, eggs enclosed in a brown flask-shaped sac close to the female's web.
Habitat and range: Rough grassland, waste land, hedgerows, and garden borders. Southern and central Europe: only near south coast in Britain.

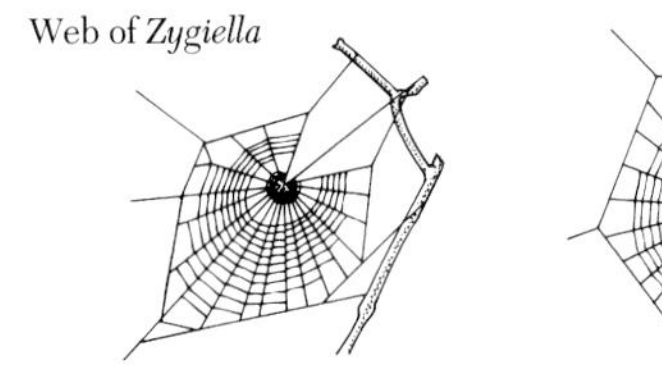
Web of *Zygiella*

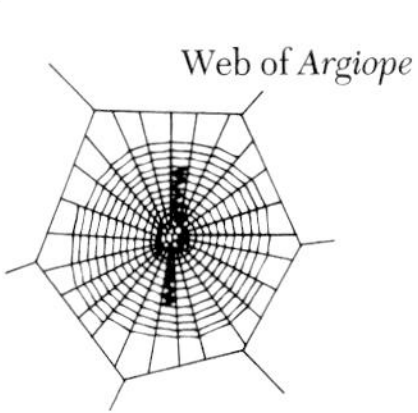
Web of *Argiope*

1

2

3

1

Amaurobius fenestralis

Description: 4–10mm: slow-moving with fairly short legs. Thorax shiny brown, often with a radiating pattern and usually contrasting sharply with the dark head. Abdomen greyish or yellowish brown, with a prominent dark mark at the front and variable dark mottling behind.
Habits: Makes a lace-like web around a crevice – usually in a wall, fence, or tree-trunk. The spider waits in the crevice for prey to arrive. The silk is not coated with gum, but it forms such a fine mesh that crawling insects get their feet firmly trapped as soon as they tread on it. New strands of silk are added haphazardly from time to time and the webs soon look rather scruffy. Adult throughout the year.
Habitat and range: Woods, waste ground, orchards, gardens, and wherever else there are trees or piles of debris. Very common on close-boarded garden fences. Throughout northern and central Europe.
Similar species: The slightly larger *A. similis* prefers somewhat drier walls and fences. The much darker *A. ferox*, has a faint skull-and-crossbone pattern on the abdomen and lives under logs and in damp sheds.

2

House Spider

Tegenaria gigantea
Description: 10–18mm, with very long legs, especially in the male. The body is yellowish to reddish brown, heavily mottled with black and with a number of clear, pale chevrons towards the rear.
Habits: Most often seen scurrying across floors at night, especially in the autumn when males are seeking mates, this is one of the spiders that build triangular cobwebs in neglected corners. The spider hides in a tubular retreat in the corner and waits for flies and other insects to become entangled in the dense mesh of the sheet. Adult all year, the females living for several years and surviving for months without food.
Habitat and range: In and around houses and other buildings, especially sheds and other out-buildings: also in open countryside, particularly in the south. Throughout Europe.
Similar species: There are several closely related species, all very similar and not easy to separate without microscopic examination.

3

Enoplognatha ovata

Description: 3–6mm. The thorax is very pale brown, while the bulbous abdomen may be cream with red bands (pictured here), entirely cream, or cream with a broad red band down the middle. There are always a number of paired black dots.
Habits: Spins a flimsy 3-dimensional web on low-growing vegetation. The outer threads are sticky and trap aphids and other small insects (larger prey, such as the bee pictured here, is unusual). The female spider is most often found in a curled-up leaf, together with her blue-green egg-sac. Adult in summer and autumn.
Habitat and range: Dense vegetation of all kinds, including shrubberies, nettle beds, and herbaceous borders. Throughout Europe.

1

2

3

1

Linyphia triangularis

Description: 4–6mm. The thorax is light brown with darker edges and a dark mark like a tuning fork in the middle. The female abdomen, markedly triangular in profile, is mainly white with a string of brown triangular marks arranged like a pagoda in the centre. There are also dark streaks at the sides. The male abdomen is much slimmer and lacks the triangles.
Habits: Makes a hammock web in bushes and hedgerows and hangs from the lower surface waiting for insects to fall on to the web – usually after bumping into the numerous threads supporting the hammock. The latter is usually flat or slightly domed, although it sags when coated with dew on autumn mornings. Adult mid-summer to late autumn.
Habitat and range: Almost anywhere with trees, shrubs, or sturdy herbaceous plants: one of the commonest spiders of garden hedges and shrubberies, often with dozens of webs on a single bush. Most of Europe.
Similar species: Several species are superficially similar, but rarely occur in gardens.

2

Wolf Spider

Pardosa amentata

Description: 5–8mm. Dark grey or brown, sometimes almost black, usually with a pale stripe down the middle of the thorax and a pale band, often broken into dots, at each side.
Habits: A fast-running, diurnal hunter, usually seen scurrying over the ground or low vegetation in search of flies and other small insects. The female carries her eggs in a lens-shaped silken parcel attached to her spinnerets, and can often be seen sun-bathing at this time. When the spiderlings hatch they climb on to the female's back and are carried around for a while, gradually dropping off to make their own way in the world. Male spiders can sometimes be seen signalling to the females by waving their black, clubbed palps. Adult in summer and autumn.
Habitat and range: Most open habitats with plenty of bare ground and some shelter and moisture: often common on garden rockeries. All Europe.

3

Nursery-web Spider

Pisaura mirabilis

Description: 10–15mm. Ground colour ranges from grey to dark brown, but is usually light brown with a pale streak on the middle of the thorax. The abdomen, always pointed at the rear, is often unmarked, but may have a dark leaf-like pattern in the centre – especially in the male.
Habits: A fast-running diurnal hunter, usually seen on low vegetation. At rest, the two front legs on each side are extended forward and each pair is held very close together. The female carries her egg-sac in her fangs, but when the eggs are about to hatch she fixes it to a plant and covers it with a tent of silk, on which she sits guard until the babies are ready to disperse. Adult in summer.
Habitat and range: Almost anywhere with plenty of low-growing vegetation: very common in nettle beds and herbaceous borders. Throughout Europe.

1

2

3

1

Woodlouse Spider

Dysdera crocota

Description: 10–15mm, with huge fangs. The thorax and legs are chestnut brown and the abdomen is flesh coloured.

Habits: A slow-moving nocturnal hunter preying entirely on woodlice, which are easily speared by the huge fangs. The spider can inflict a painful bite if handled. It hides under logs and stones by day. Adult all year.

Habitat and range: Woods, hedgerows, waste land, gardens, and anywhere else damp enough to shelter woodlice. Often found in log-piles and compost heaps in the garden. Southern and central Europe: quite common in southern Britain but rare in the north.

Similar species: D. erythrina is a little smaller but otherwise almost identical. *Harpactea rubicunda*, rare in the British Isles, is similar in colour but has normal-sized fangs.

2

Zebra Spider

Salticus scenicus

Description: 5–7mm, with a variable pattern of black and white bands on the abdomen. The head is square in front, with two very large eyes in the centre. The male has greatly enlarged fangs.

Habits: This is one of the jumping spiders that stalk their prey and then leap on to it, pinning it down with the enlarged front legs. It is active by day and its large eyes enable it to pin-point the position of its victims very clearly. Small flies are its main victims. Adult throughout the summer.

Habitat and range: Most common on and around buildings, especially on old walls and fences where it is well camouflaged among the lichens. Also found on rocks and tree trunks. Throughout Europe.

Similar species: There are several similar species, but they are not often found in gardens.

3

White Death

Misumena vatia

Description: Male 3–4mm: female 9–12mm. The female, which is the only sex commonly noticed, is white, yellow, or pale green, sometimes with red spots or stripes on the sides of the very plump abdomen. The two front pairs of legs are much larger than the other two pairs.

Habits: This crab spider sits motionless in flowers – usually white or yellow ones – and ambushes insects that come to feed. It can tackle prey much larger than itself, including bees and butterflies. The victims are bitten just behind the head as they feed and are paralysed immediately. The spider is able to change it colour slowly from white to yellow and vice-versa to match the flowers. Adult throughout the summer.

Habitat and range: Almost any flower-rich habitat, including roadside verges and herbaceous borders. Throughout Europe.

1

2

3

Harvestmen

The harvestmen, sometimes called harvest spiders, are only distantly related to spiders. Their globular bodies are not obviously divided into two sections and they have just two eyes, perched on a turret a little way behind the front of the body. Most have very long legs, with the second pair always the longest. Harvestmen feed mainly on small invertebrates, both living and dead. They produce neither venom nor silk. Most species pass the winter as eggs and mature in late summer – hence the common name.

1 *Opilio parietinus*

Description: 5–9mm. Brown or greyish brown, crossed by darker bands and often bearing a pale stripe running along the back from the eye-turret to the rear of the body. There may be a vague, dark saddle-like mark in the middle, although this is usually absent in the male, and the whole upper surface is crossed by bands of pale, dark-tipped tubercles. The underside is off-white with brown spots.
Habitat and range: Tree trunks, bushes, and rough grass, but most common around human habitation – on walls and fences in parks and gardens. Most of Europe except the far north.

2 *Odiellus spinosus*

Description: 6–11mm and rather flat. Greyish brown, with a black-edged, rectangular 'saddle' in the middle of the back, sharply truncated by a black line at the rear. There is a trident of three sturdy spines on the front edge of the body, linked to the eye-turret by a pale stripe that narrows towards the rear. The legs are shorter than in most other harvestmen of similar size.
Habitat and range: Strongly associated with human habitation; occurs in parks and gardens and on waste ground. Usually keeps close to the ground and can often be found in rough grass and under pinks and other sprawling garden plants: sometimes climbs walls and tree trunks. Southern and central Europe: confined to southern and central England in Britain.
Similar species: Oligolophus tridens is superficially similar but is smaller and lacks the white streak from the trident to the eye-turret.

3 *Leiobunum rotundum*

Description: Male 3–4mm: female 4–7mm. The male's body is rusty-brown and almost circular, with no obvious markings apart from the black eye-turret. The female's body is oval and paler brown with a dark, more or less rectangular 'saddle'. Both sexes have extremely long, hair-like black legs.
Habitat and range: Trees, shrubs, and dense vegetation of all kinds, including nettle beds and garden borders. Often seen resting on walls by day, with their legs spread widely and almost invisible. Commonly enters porches and other buildings where lights are left on for much of the night. Most of Europe except the far south and south.

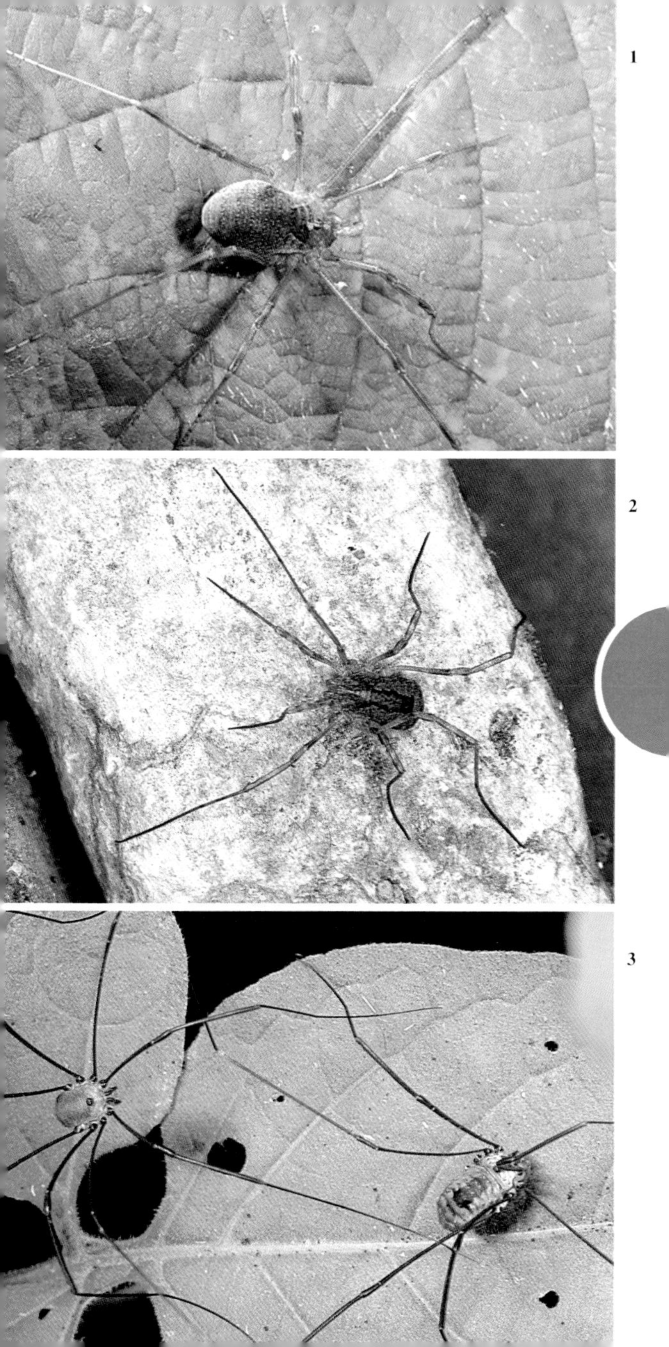

1
2
3

Woodlice

Woodlice are land-living crustaceans and, although some species can withstand drier conditions than others, they have not completely shrugged off their watery ancestry and are absent from the driest habitats. Their oval bodies consist of two main regions – a thorax or pereion with seven pairs of legs, and a slightly narrower pleon behind. All feed mainly on decaying vegetation and help to recycle nutrients and, although they sometimes nibble living plants, they probably do more good than harm. The animals moult in two stages, shedding the skin from the rear half first and from the front half a few days later. The freshly-moulted region is white, so it is not uncommon to find a woodlouse that is half grey and half white.

1

Oniscus asellus

Description: 15mm. The body has a smooth outline, with no obvious junction between thorax and pleon. Shiny grey with pale blotches, usually cream or yellowish, at the sides. Flagellum of antenna (beyond the longest section) has three segments.
Habitat and range: The commonest woodlouse in most areas, it is abundant in gardens, especially under logs and in the compost heap. Most of Europe.
Similar species: Porcellio spinicornis has a mottled brown body with yellow spots and a dark line down the centre. The head is black with prominent tubercles, and the flagellum has only two segments.

2

Philoscia muscorum

Description: 11mm. Usually brown, but sometimes red or yellowish, and always with a darker stripe along the middle of the back. There is an abrupt junction between the thorax and the much narrower pleon. The flagellum of the antenna has three segments.
Habitat and range: Turf, leaf litter, and vegetable debris of all kinds: common in the compost heap and also in quite dry hedge-bottoms. Throughout the British Isles and most of Europe, but uncommon in Scotland and other northern areas.

Dorsal view of *Oniscus asellus*

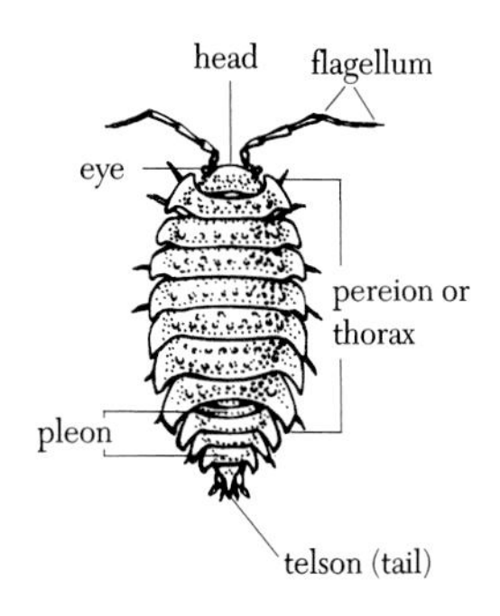

1

2

1

Androniscus dentiger

Description: 6mm. The body ranges from white to deep pink, usually with a dark central stripe, and **is** covered with prominent, spined tubercles. There is a fairly abrupt junction between the thorax and the pleon and the flagellum of the antenna is not obviously segmented.
Habitat and range: Compost heaps and other vegetable debris: also in damp cellars. Most of Europe except the far north: rare in Scotland.

2

Porcellio scaber

Description: 17mm. Generally dull grey, with numerous tubercules, but sometimes yellowish or orange with grey or black spots. The front of the head is bluntly pointed and the flagellum of the antenna has two segments. There is a smooth transition between thorax and pleon.
Habitat and range: A very common woodlouse, often cohabiting with *Oniscus* under logs and in compost heaps, but able to tolerate slightly drier habitats and commonly found on tree trunks and walls, where it feeds on algae at night. Throughout Europe.

3

Armadillidium vulgare

Description: 18mm. This is one of the pill woodlice or pill-bugs that are able to roll into balls when disturbed. The body is smooth and shiny and the upper surface is strongly domed. It is usually slate-grey, often with a bluish tinge, although it is sometimes brown or even yellowish. The sides of the thorax are almost parallel, with a smooth transition between them and the pleon. When rolled-up, the small plates of the pleon are clearly visible.
Habitat and range: Tolerant of drier habitats than most woodlice, partly because of its ability to roll up, this species is largely restricted to calcarous soils. In gardens, it is most often found at the foot of old walls. It is by far the commonest and most widely-distributed of the pill-bugs, although scarce in northern Britain and other northern regions.
Similar species: There are several very similar species, but they are much less common. The pill millipede (p.182) is much blacker and even shinier, and has many more legs. It lacks the numerous small plates at the rear.

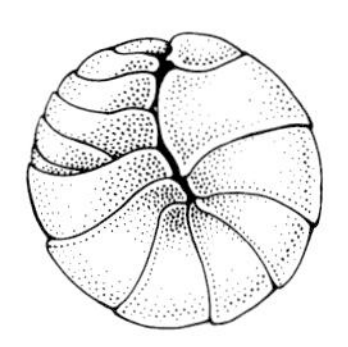

rolled up Pill Woodlouse

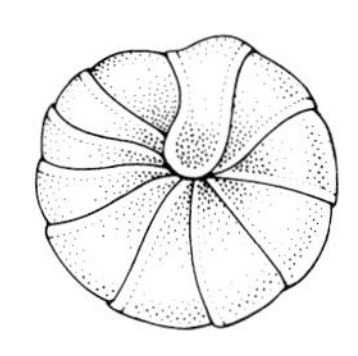

rolled up Pill Millipede

1
2
3

Centipedes and Millipedes

Centipedes and millipedes commonly live under the same logs and stones and, because both groups have many legs, they are frequently mistaken for each other. But they are not closely related. Centipedes (Class Chilopoda) have one pair of legs on each body segment, while millipedes (Class Diplopoda) have two pairs of legs on each body segment. Centipedes are carnivorous creatures, with poison claws just behind the head, and they feed on a wide range of other invertebrates, including other centipedes. Millipedes are herbivorous animals, feeding mainly on dead plant matter, although some damage living plants. The two classes are sometimes lumped together in a single group known as myriapods.

1 *Haplophilus subterraneus*

Description: Up to 70mm, with from 77–83 pairs of legs. Yellow or pale brown, tapering strongly towards the slightly darker head.
Habits: A blind, burrowing centipede, but more often found under stones and in leaf litter than truly in the soil. Essentially carnivorous, but sometimes causes damage by nibbling plant roots. May glow when disturbed at night.
Habitat and range: A wide range of soil habitats, but especially under grassland and arable crops: quite common in gardens and orchards. Most of Europe, but only in association with man in Scandinavia.

2 *Geophilus carpophagus*

Description: Up to about 60mm, although usually no more than about 40mm, with from 45–55 pairs of legs. Chestnut brown, with no eyes.
Habits: Essentially a soil-dwelling centipede, often found lurking under stones. Glows strongly when disturbed at night and called a glow-worm in some parts of the country.
Habitat and range: Primarily a woodland creature, but it occurs in the soil in many gardens and orchards and can also be found in cellars and damp out-buildings. Throughout Europe.

3 *Necrophloeophagus longicornis*

Description: Up to 45mm, although few specimens exceed 30mm. Bright yellow with a darker head and 49 or 51 pairs of legs. The antennae are considerably longer than those of other soil-dwelling centipedes.
Habits: A blind, burrowing species, often found quite deep in the soil. As in the other burrowing centipedes, its body is amazingly flexible.
Habitat and range: Almost any kind of soil, from the upper reaches of the seashore to alpine pastures: very common in gardens, even those in the middle of towns. Throughout Europe.

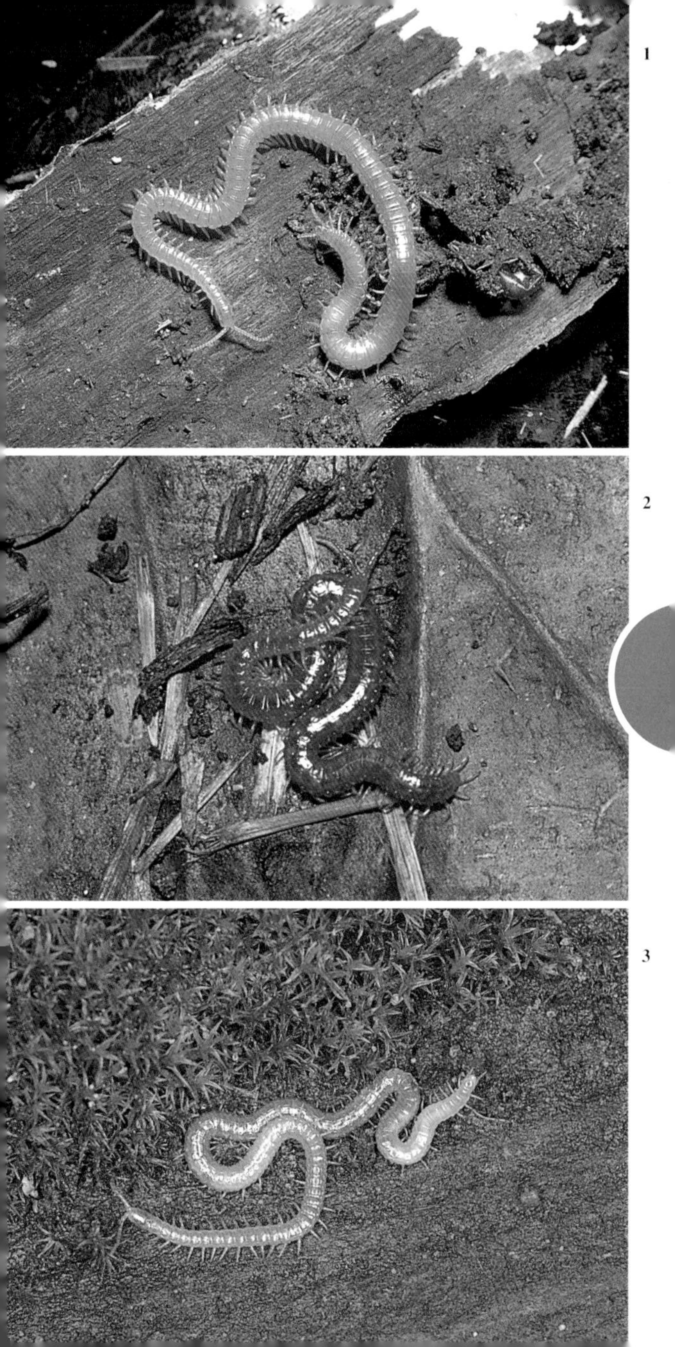
1
2
3

1

Cryptops hortensis

Description: Up to 30mm, with 21 pairs of legs – of which the last pair are very long and thick. Pale, shiny brown, with no eyes.
Habits: A very active centipede, hiding under logs and stones or loose bark by day and hunting by night.
Habitat and range: Although it occurs widely in woods and grasslands, this centipede seems to have a particular liking for gardens, where it can be found under logs and flower-pots during the day-time. Most of Europe, but confined to human settlements in Scotland and Scandinavia.

2

Lithobius forficatus

Description: Up to 30mm long and 4mm wide. Shiny brown, with 15 pairs of legs when mature. The legs are much longer than those of the burrowing centipedes and the hind legs are particularly long. The head capsule is much rounder than that of the burrowing centipedes and carries a number of simple eyes on each side.
Habits: A very active centipede, hiding under logs and stones by day and hunting at night. Insects, worms, slugs, and other centipedes all succumb to its venomous claws. Unlike those of the burrowing centipedes, the young do not have the full complement of legs. They have only seven pairs when they leave their eggs and they add one or more pairs at each moult until they have all fifteen pairs.
Habitat and range: Occurs almost everywhere, from the seashore to moors and mountain-tops. Very common in gardens, where it often invades sheds and greenhouses. Throughout Europe.
Similar species: There are many similar species, but only the smaller and paler *L. duboscqui* is at all common in gardens.

3

House Centipede

Scutigera coleoptrata
Description: Up to 30mm long and 5mm wide, with 15 pairs of legs and a pair of very long antennae. Dull brown with dark stripes and yellow spots, and a large dark eye on each side of the head. The 15 pairs of legs are very much longer than those of our other centipedes, the last pair being as long as the body and so slender that they are easily confused with the antennae.
Habits: An extremely fast runner, active by night and feeding on flies, spiders, and other small invertebrates.
Habitat and range: Usually found in and around buildings, especially on walls: often called the house centipede. A native of southern Europe; well established in the Channel Islands, but otherwise only a sporadic visitor to the British Isles – probably arriving in tomatoes and other produce.

1
2
3

1 Pill Millipede

Glomeris marginata

Description: Up to 20mm long and 8mm wide – much wider than the other groups of millipedes. Shiny black or deep brown, with 17–19 pairs of legs and a broad, almost semi-circular plate at the rear.
Habits: Plays an important role in breaking down dead vegetation in some places. May also nibble plant stems. Rolls into a ball when disturbed.
Habitat and range: Mainly in leaf litter and turf: in the garden it is most often found in hedge bottoms and in debris at the base of walls. Withstands drier conditions than most other millipedes. Most of Europe.
Similar species: The Pill Woodlouse (p.176) is greyer and has only 7 pairs of legs, and lacks the large semi-circular plate at the rear.

2 Flat-backed Millipede

Polydesmus angustus

Description: Up to 25mm long and 4mm wide, with 37 pairs of legs and flat, wing-like extensions to the body segments. The upper surface is heavily sculptured.
Habits: Feeds mainly on decaying vegetation, but also nibbles plant roots and sometimes damages strawberries and other soft fruit.
Habitat and range: Leaf litter, turf, and soils with a high organic content: often abundant in garden compost heaps. Most of Europe.
Similar species: There are several other flat-backed millipedes, all distinguished from centipedes by the two pairs of legs on each segment.

3 Snake Millipede

Tachypodoiulus niger

Description: Up to 50mm long and 4mm in diameter: cylindrical, shiny black, and tapering slightly towards each end.
Habits: Lives in the surface layers of the soil, in leaf litter, and under loose bark. Feeds on both living and dead plant matter and is strongly attracted to soft fruit: often climbs raspberry canes and other bushes to feed at night. Coils up like a flat spring when disturbed.
Habitat and range: Most well-vegetated habitats, including garden hedges and herbaceous borders. Most of Europe.
Similar species: Several other snake millipedes occur in the garden, although most are smaller and less addicted to climbing.

4 Spotted Snake Millipede

Blaniulus guttulatus

Description: About 15mm long and very slender. Easily recognised by its pale colour and the red spots along the sides – these are glands that release pungent protective fluids.
Habits: A burrowing millipede, feeding mainly on dead organic matter but often turning its attention to plant roots and tubers in dry conditions: it can cause severe damage to potatoes, sugar beet, and other crops.
Habitat and range: Cultivated soils, especially the heavier and damper ones. Most of Europe.

1

2

3/4

1

Shelled Slug

Testacella haliotidea

Description: Up to 12cm, creamy white or pale yellow with a small flat shell, not unlike a finger nail, perched on the rear. A dark branching line runs forward along each side of the body from the shell and the sole is usually white.

Food and habits: This slug lives in the soil and feeds almost entirely on earthworms: when fully extended, it is very slender and can follow the worms along their tunnels. It is rarely seen above ground, although it can sometimes be found resting under paving slabs laid in the garden. At rest it is short and plump and almost pear-shaped.

Habitat and range: Confined largely to well-manured, well-drained cultivated soils, mainly in parks and gardens: sometimes in compost heaps. Southern and Central Europe: most common in the west.

Similar species: T. maugei is darker, with a much larger shell and often with a pinkish foot-fringe. *T. scutulum* is usually yellow with an orange sole and fringe.

2

Great Grey Slug

Limax maximus

Description: Up to 20cm: pale brown or grey, heavily marbled with dark spots, those on the rear half of the body often linked to form stripes. The head and tentacles are reddish brown. There is a short keel on the rear end of the body.

Food and habits: Eats fungi and rotting vegetation and is harmless in the garden. The mating behaviour is truly amazing. After meeting on the ground, two slugs climb a wall or fence or some other vertical surface and then lower themselves on a tough rope of mucus. Mating finally takes place with the two animals dangling at the end of this rope, with their shiny white genitalia entwined as shown in the picture. The whole process may take an hour or more. One slug then drops to the ground and the other usually climbs back up the rope and eats it as it goes. Both partners then go off and lay eggs because, in common with our other garden slugs and snails and the earthworms, the animals are hermaphrodite and each individual contains both male and female organs (see p.196).

Habitat and range: Woods, hedgerows, and gardens – especially in and around compost heaps. Much of Europe except the far north: mainly around human habitation in northern and eastern parts of the range.

3

Yellow Slug

Limax flavus

Description: Up to 10cm: yellow with pale grey spots, blue tentacles, and yellow body slime. There is a short keel at the rear of the body.

Food and habits: Eats fungi and rotting vegetation: harmless in the garden.

Habitat and range: Gardens, cellars, damp sheds, etc. Rarely found away from human habitation. Western Europe, including southern Scandinavia.

1

2/3

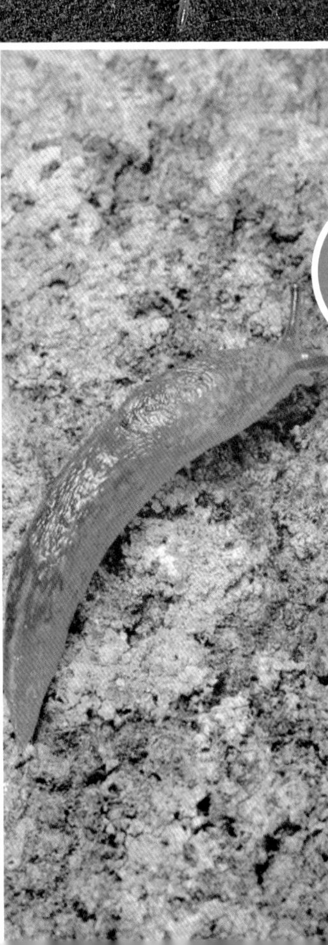

3

1

Netted Slug

Deroceras reticulatum

Description: Up to 5cm, cream to light brown or dark grey, usually with darker flecks and a prominent mosaic of rectangular tubercles. There is a short keel at the rear. The slug exudes a white mucus when disturbed.
Food and habits: A major garden pest: attacks a wide range of plants and completely destroys newly planted seedlings. Generally rests among leaves, often right inside lettuces and cabbages: rarely in the soil.
Habitat and range: Hedgerows, gardens, arable land, and rough grassland. Throughout Europe, although mainly in and around human habitation in the far north. One of Europe's commonest slugs.
Similar species: D. sturanyi lacks dark flecks; sole is pale with a dark band down the middle. Occurs mainly in Central and Eastern Europe but is spreading westwards – probably with the horticultural trade. *D. caruanae*, greyish sole and colourless body mucus, occurs mainly in Western Europe.

2

Sowerby's Slug

Milax sowerbyi

Description: Up to 7.5cm: greyish brown, heavily speckled with black and bearing an orange or yellowish keel running from the mantle to the end of the body. The sole is pale and the body mucus is yellowish.
Food and habits Feeds mainly on roots and tubers and spends most of its time in the soil: a serious pest of potatoes and other root crops. Contracts to a dome when disturbed.
Habitat and range: Mainly in gardens and arable fields. Southern and Central Europe: most common in the west.
Similar species: Budapest Slug (below) has colourless mucus. Smooth Jet Slug (*Milax gagates*) has a dark keel and colourless mucus.

3

Budapest Slug

Milax budapestensis

Description: Up to 6cm: greyish brown to dark grey with darker spots and a yellow or orange keel. The sole is dirty yellow with a dark band down the centre and the body mucus is colourless.
Food and habits: Feeds mainly on roots and tubers and spends most of its time in the soil. In common with other *Milax* species, it is a serious pest of root crops. Rather slender when extended, it adopts a characteristic 'C' or comma-shape when it contracts.
Habitat and range: Confined largely to gardens and other cultivated land. Widely distributed in Central Europe, but most common in the west, especially in the British Isles.
Similar species: Sowerby's Slug (above) has yellow mucus.

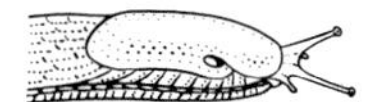
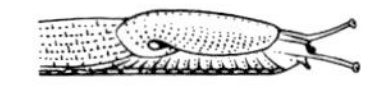

Front ends of round-backed (left) and keeled (right) slugs showing different positions of the breathing pore

1

2

3

1

Large Black Slug

Arion ater

Description: Up to about 20cm when fully extended, but usually 10–15cm. Ranges from jet black, through chestnut and orange (1b), to pale grey or creamy white. Brown and cream forms always have an orange fringe to the sole. There is no keel on the back, but the body is covered with elongated tubercles. When disturbed, the slug contracts to a hemisphere and often sways from side to side. Its mucus is extremely sticky.
Food and habits: An omnivorous species, eating carrion and dung as well as vegetable matter. It prefers rotting vegetation to living plants and rarely does much harm in the garden. Largely nocturnal, but large numbers may gather on freshly-mown roadside verges to feed on the cut grass after daytime rain. The clusters of pearly round eggs (1a), about 5mm in diameter, are often dug up in the garden and the compost heap.
Habitat and range: Almost any well-vegetated habitat. Most of Europe, including Iceland. The black form is most common in northern areas and the paler ones are more common in the south.

2

Garden Slug

Arion hortensis

Description: Up to 4cm: bluish-black on the back, with paler flanks and a conspicuous dark band on each side. Its tentacles are reddish brown and the sole is orange. There is no keel. Its mucus is orange or yellow.
Food and habits: Attacks almost any herbaceous plant, above or below ground, and is a serious garden pest: often attacks potato tubers in company with Sowerby's Slug (p.186).
Habitat and range: Woods and hedgerows, but most common on cultivated land. Most of Europe, but absent from the far north.
Similar species: A. distinctus and the much rarer *A. owenii* are hard to distinguish with certainty and were once regarded as races of *hortensis*. *A. distinctus* is often quite yellow below the dark stripe on each flank. It has dark tentacles, often with a bluish tinge. *A. owenii* is browner, with reddish brown or purplish tentacles and a yellower sole. Bourguignat's Slug (*A. fasciatus*) has a plain white sole and colourless mucus. It feeds mainly on fungi and rotting leaves.

3

Dusky Slug

Arion subfuscus

Description: Up to 7cm: yellowish brown to chestnut, sometimes greyish, with a darker line on each side and no keel. The sole is pale yellow. The body mucus is bright yellow or orange, staining anything it touches.
Food and habits: Feeds mainly on fungi and decaying matter and does little harm in the garden. Unlike the other *Arion* species, it cannot contract into a hemisphere.
Habitat and range: Woods, hedgerows, gardens, grassland, and waste ground. Throughout Europe.

1/1a
1b
2/3

1 Garlic Glass Snail

Oxychilus alliarius

Description: Shell up to 7mm across and nearly flat: shiny and translucent. The horny shell is actually pale yellow or greenish, but the snail's slate-grey body gives it a dark brown appearance in life. The snail emits a strong smell of garlic when handled.

Food and habits: Feeds mainly on fungi and rotting vegetation. Mainly ground-living, but sometimes climbs damp walls and trees at night.

Habitat and range: Abundant in leaf litter, especially under logs and stones. A common inhabitant of the garden compost heap. Confined mainly to the British Isles and neighbouring countries bordering the Atlantic or North Sea, including Iceland.

Similar species: Cellar Glass Snail (*O. cellarius*) has a paler body and its shell is usually 10–12mm across. It lives in leaf litter, compost heaps, and many other damp places, including cellars. It feeds on fungi and small animals, including worms and insect grubs.

2 Draparnaud's Snail

Oxychilus draparnaudi

Description: Shell up to 15mm across, yellowish brown and less translucent than that of other *Oxychilus* species. The opening is about half of the total width of the shell. The body is deep blue-black.

Food and habits: Largely carnivorous, feeding on small earthworms and insect larvae. Mainly ground-living.

Habitat and range: In leaf litter and other debris, mainly in gardens, where it is a common inhabitant of the compost heap: often shelters under logs and stones. Western Europe, but rare and usually confined to gardens and greenhouses in the north.

3 Strawberry Snail

Trichia striolata

Description: Shell up to 14mm across, dirty yellow to reddish brown or purple: a flattened cone with a blunt spire and rough growth ridges running across each whorl. There is a prominent white ring just inside the mouth of the shell. The body is dark grey.

Food and habits: Browses on a wide range of low-growing plants, damaging strawberry fruits, lettuces, and many other garden plants. Mainly nocturnal, but often active after rain in the daytime. It hides under the vegetation by day and may be found in large numbers under pinks and similar mat-forming plants.

Habitat and range: Most well-vegetated habitats, especially in low-lying areas with plenty of moisture. The British Isles and neighbouring countries, extending in a narrow belt across Germany to Hungary.

Similar species: Lapidary Snail has a sharp keel around the edge.

1

1a

2

1 Rounded Snail

Discus rotundatus

Description: Shell up to 7mm across: rather flat, with prominent ribs and conspicuous rust-coloured stripes on a yellowish background. The underside has a very wide hollow – the umbilicus – occupying about one-third of the shell's diameter.

Food and habits: Feeds on a wide range of decaying matter and fungi.

Habitat and range: Lives in damp vegetation and leaf litter: sometimes abundant in garden compost heaps. Most of Europe, but confined largely to gardens in the north.

Similar species: The Lapidary Snail has a similar pattern but is much larger when mature and has a sharp keel running all the way round the outer edge.

2 Roman Snail

Helix pomatia

Description: Shell up to 50mm across: thick and globular with a large, rounded opening and thick white lip. Usually creamy yellow with coarse growth lines and indistinct spiral brown bands. Body is yellowish grey.

Food and habits: A vegetarian, browsing on a wide range of low-growing plants and sometimes climbing shrubs to reach tender leaves. Becomes dormant in the coldest months, burrowing in turf or leaf litter and closing the mouth of its shell with a sheet of lime-rich mucus that dries to form a tough plate called an epiphragm.

Habitat and range: Woods, hedgerows, and rough grassland on lime-rich soils: not common in gardens other than in the south, where it is also a vineyard pest. Native to Southern and Central Europe, including south-east England, it is widely reared or collected for food and has been introduced to several other areas. It is also called the edible snail.

Similar species: This is the largest snail in most parts of Europe, but there are several similar ones in the south-east.

3 Garden Snail

Helix aspersa

Description: Shell 25–40mm across: globular, with a large, rounded mouth surrounded by a thick white lip. It is essentially yellowish brown with up to five dark brown spiral bands, but the latter are usually broken up to give the shell a mottled appearance.

Food and habits: Few low-growing plants escape the attentions of this abundant pest, which can do considerable damage to strawberries, courgettes, newly planted seedlings, and many other garden plants. Active mainly at night, it often roosts communally by day at the bottoms of walls and in up-turned flower-pots. Mating pairs are not uncommonly seen early in the morning, with their white genitalia still locked together. The snails become dormant in winter, and also in very hot or dry weather, having sealed their shells with quick-drying mucus like the Roman Snail (see above).

Habitat and range: Woods, hedgerows, and waste ground, but most often in parks and gardens. Southern and Central Europe. Widely reared for food.

1

2

3

1 Kentish Snail

Monacha cantiana

Description: Shell up to 20mm across, more or less top-shaped with a very round mouth and a conspicuous white rib just inside it. Dirty white, usually tinged with reddish brown near the mouth. The body is pale brown and may show through the upper parts of the shell as a lead-coloured tinge.
Food and habits: Feeds mainly on decaying vegetation and unlikely to do any harm in the garden. Active at night or after daytime rain. Large numbers can be seen feeding on heaps of cut grass on roadside verges.
Habitat and range: Grassy places of all kinds, including orchards and hedge bottoms – from where it may spread into the herbaceous border: not uncommonly shelters under aubretias and other sprawling rockery plants. Widely distributed in Southern and Central Europe, mainly in the west.

2 White-lipped Banded Snail

Cepaea hortensis

Description: Shell up to 20mm across: top-shaped and shiny, with an oval mouth normally surrounded by a white lip. The ground colour is usually yellow and there are typically five dark spiral bands, although many snails have fewer bands and some have no banding at all. The ground colour is sometimes pink or brown. The body is greenish grey, with a yellowish patch at the rear.
Food and habits: Eats grass and other low-growing plants, including lettuces and other tender garden species. Active at night or after daytime rain.
Habitat and range: Well vegetated habitats of all kinds, especially hedgerows and rough grassland. Most of Europe, including Iceland, but absent from the far north.
Similar species: Brown-lipped Banded Snail (below) is extremely similar apart from its brown lip.

3 Brown-lipped Banded Snail

Cepaea nemoralis

Description: Shell 20–30mm across (slightly larger than White-lipped Banded Snail), usually with a brown lip but otherwise exhibiting the same colours and patterns as the previous species. The body is dark grey with little trace of yellow at the rear.
Food and habits: Eats grass and other herbaceous plants, often mingling with the previous species.
Habitat and range: Wood, hedgerows, and other rough vegetation: probably a little less common in gardens than the previous species. Southern and Central Europe – not extending as far north as *hortensis*.
Similar species: White-lipped Garden Snail (above) usually has a white lip, but this is not always a reliable feature: in some areas *hortensis* snails may have brown lips and some *nemoralis* snails have white lips!

1

2

3

1

Thunderworm

Mermis nigrescens

Description: Up to 50cm, resembling a piece of brown or white cotton. Thunderworms are unrelated to the earthworms and their bodies are not divided into rings or segments.

Habits The worm spends most of its time in the soil and is normally seen above ground only after rain, when it twines itself around low-growing plants. Females lay their eggs on the plants and the eggs are swallowed by leaf-eating beetles or other suitable insects. The eggs hatch inside the insect's body and the young worms absorb food from the insect's body fluids. When the worms mature they leave to live freely in the soil. The hosts are severely weakened but not necessarily killed.

Habitat and range: Soils of all kinds. Probably throughout Europe, but most common in the south, where there is a much denser population of host insects.

2

Common Earthworm

Lumbricus terrestris

Description: Up to 30cm: bright pink to reddish brown with a tinge of violet. There are about 150 segments and the orange-red clitellum or saddle usually covers segments 32–37.

Food and habits: Earthworms feed mainly by swallowing soil and debris and digesting any organic material in it. Their burrows play a major role in aerating and draining the soil. Undigested material passes out in the form of worm-casts, sometimes deposited on the surface but more often in the burrows. On warm nights the worms may surface and pull dead leaves into their burrows. Mating also takes place on the surface. The two worms lie side by side, although they always leave their rear ends anchored in their burrows for a quick retreat, and each individual gives and receives sperm. Each one then goes off to lay its own eggs (see p.184). The worms are active throughout the year, but they burrow deeply and become dormant in very cold or dry weather.

Habitat and range: Any soil that is not too wet or too acidic: most common under lawns and other undisturbed grassland. Most of Europe.

Similar species: Several other species live in our garden soil. They may differ in size and colour and in the position of the clitellum, but they all behave in the same way.

3

Brandling Worm

Eisenia foetida

Description: Up to 12cm: purplish brown or red with conspicuous orange bands. The worm gives out a smelly yellow fluid when handled, but despite this it is a popular bait with anglers.

Habitat and range: Requires soils with a high organic content. Abundant in compost heaps and manure heaps and in recently manured ground: also thrives under the bark of fallen and rotting tree trunks and logs. Some people culture this worm in bins, where it quickly converts organic kitchen waste into compost. Most of Europe except the far north.

1

2

3

Mosses and Liverworts

Mosses are flowerless, mat-forming or cushion-forming plants with slender stems, small leaves, and no true roots. They reproduce by scattering dust-like spores from capsules growing from the tips of the stems. But capsules do not develop until male and female cells produced on the stems have paired up. This can happen only when the plants are wet, so mosses flourish best in damp places. The spore capsules have pointed hoods at first, but these fall off to release the ripe spores, which then grow into new mosses. Liverworts have simple round capsules without hoods. Most of them grow just like mosses but the commonest garden liverworts look more like seaweeds, with no separation into stems and leaves.

1 Wall Screw Moss

Tortula muralis

Description: Forms neat, bright green cushions up to 5cm across and 2cm high. Each tiny leaf ends in a little white hair, and when the leaves curl up in dry weather the whole cushion may appear quite grey. The slender, cylindrical spore capsules are carried vertically on orange or reddish stems in the spring.

Habitat and range: Grows on rocks and the tops of walls, where it is one of the commonest species, and also in cracks in concrete paths. Throughout Europe.

2 Grey Cushion Moss

Grimmia pulvinata

Description: Forms tight greyish green cushions, up to 3cm high and often more or less hemispherical. The leaves are dark green, but each ends in a long grey hair and the hairs give the whole cushion its greyish colour. The oval, green spore capsules are carried on curved stalks and are hidden among the leaves at first, but the stalks straighten out when the spores are ripe and carry the capsules well above the cushions so that the spores can be scattered in the breeze.

Habitat and range: Rocks and walls, often growing with Wall Screw Moss. Also on roofs and concrete paths. Throughout Europe, but most common in lowland areas and rarely found above 1000m.

3 Matted Thread Moss

Bryum capillare

Description: Forms dense, bright green mats or cushions up to 5 or 6cm high. Each tongue-shaped leaf ends in a greenish hair, so the moss never looks grey even when the leaves curl up in dry weather. The spore capsules are large and pear-shaped, green at first but becoming orange or reddish brown as they ripen.

Habitat and range: Rocks, old walls, roofs, tree trunks, and fences, wherever there is a reasonable accumulation of soil or other debris. Throughout Europe.

1

2

3

1

Silvery Thread Moss

Bryum argenteum

Description: Forms dense, silvery mats or cushions up to 1.5cm high. The leaves are very small and each has a silvery tip. They are packed so tightly around the stems that the shoots look like short pieces of cord. The small, oval spore capsules hang from short stalks and are bright red when ripe.
Habitat and range: Crevices in rocks, walls, and paths, including town pavements. Common on rockeries and even colonises stone window sills. Throughout Europe and most of the world.

2

Silky Wall Feather Moss

Camptothecium sericeum

Description: Forms extensive golden green mats, composed of long, branched creeping shoots from which densely packed upright shoots grow like the pile of a carpet. The creeping shoots have sharply pointed tips and the golden tinge is particularly noticeable here. The shoots cling tightly to the substrate in damp weather, but they curl up in dry conditions, as shown here, and can then withstand quite long periods without water. The leaves are triangular, tapering to a long point. Spore capsules are more or less cylindrical and 2–3mm long, and are carried vertically on orange stalks.
Habitat and range: Rocks and walls, especially on limestone: often smothers the stones of neglected rockeries and also grows on the bases of tree trunks. Throughout Europe.

3

Rough-stalked Feather Moss

Brachythecium rutabulum

Description: Forms extensive, loose and rather shiny yellowish green carpets. The much-branched, creeping stems carry broad, oval leaves that taper to a sharp point. The spore capsules, up to 3.5mm long and slightly curved, are carried horizontally on stalks that are slightly rough to the touch.
Habitat and range: Grows on rocks and decaying wood, on the bases of tree trunks, and amongst grass on damp soils. Often invades neglected rockeries and unkempt lawns, especially around trees. Throughout lowland Europe. *B. velutinum* is very similar, but has narrower leaves with straighter margins.

4

Common Cord Moss

Funaria hygrometrica

Description: Forms pale green cushions or extensive mats, rarely more than 1cm high. Each patch contains numerous unbranched or sparsely-branched, vertical stems, each with a few oval leaves. The drooping, pear-shaped spore capsules, green at first but orange or brown when ripe, are carried on flame-coloured stalks.
Habitat and range: Grows on bare soil in a variety of situations. In the garden it is most often seen on the site of a bonfire, on the edges of paths, or smothering the soil in neglected flower-pots. In the past it used to be abundant on old cinder paths. Throughout Europe and most of the world.

1

2

3/4

1 Crescent-cup Liverwort

Lunularia cruciata

Description: The sprawling, leathery green lobes of this plant are easily recognised by the crescent-shaped cups near the tips. These cups contain small detachable buds called gemmae, that are washed out by rain drops and then grow immediately into new plants. This liverwort hardly ever produces spore capsules.

Habitat and range: The plant can be found on river banks and damp woodland rides but is most common in gardens. It is abundant on damp paths and walls and in greenhouses, where it enjoys the warmth and often covers the ground and the soil in undisturbed flower-pots. Most of Europe, although rare in the north.

2 Common Liverwort

Marchantia polymorpha

Description: The sprawling, branched green lobes of this liverwort can be recognised immediately by the crown-shaped gemma cups near the tips. These cups contain detachable buds that are scattered by rain drops (see above). The plant also produces stalked, umbrella-shaped structures that bear the sex organs. Spore capsules develop after fertilisation on the underside of the domed, star-like female umbrellas pictured here. Male umbrellas, produced on separate plants, are flat-topped.

Habitat and range: Damp soil in various habitats, including river banks woodland rides, and shady garden paths. Commonly smothers the soil in damp flower-pots, in both the greenhouse and the open garden. Most of Europe.

3 Field Horsetail

Equisetum arvense

Description: Distantly related to the ferns, this horsetail produces two very different kinds of shoots. The unbranched brown and cream fertile shoots are the first to appear in the spring. Each one is tipped by a more or less oval brown cone that opens to release large numbers of greenish spores, which develop in much the same way as fern spores (see p.204). The fertile shoots wither and die after scattering their spores, but by this time the sterile green shoots have appeared. These bear whorls of slender branches and resemble miniature Christmas trees (3a). They are clothed with silica crystals, which make them rough to the touch and which led to their being used for scouring cooking pots in the past. The shoots die down in the autumn, but the extensive system of roots and rhizomes remains alive in the soil, often at considerable depth.

Habitat and range: Roadsides, river banks, and disturbed land of all kinds. Often invades gardens and other cultivated land, where it is difficult to eliminate because of its extensive underground system. Throughout Europe, including the Arctic.

1

2

3/3a

Ferns

Ferns are flowerless plants that reproduce by scattering spores which are carried on the undersides of the (usually) deeply divided leaves or fronds. The spores germinate in damp conditions to produce minute green plates called prothalli. Male and female cells develop on the prothalli and, after fertilisation, the female cells grow into new fern plants. Moisture is vital for the reproductive process and ferns therefore flourish best in damp places. All the ferns described here remain green through the winter.

1 Rusty-back Fern

Ceterach officinarum

Description: A densely tufted fern with dull green fronds up to about 15cm long. Each side of the frond bears a number of broad-based oval, lobes, arranged alternately, giving the frond a distinctly wavy outline. The underside of the frond has a dense coating of brown scales, from which the name is derived. The spore capsules are hidden among the scales. In dry weather the fronds curl up and may appear dead, but the scales prevent excessive water loss and the fronds quickly regain their normal appearance after rain.
Habitat and range: Crevices in lime-rich rocks, and especially in the mortar of old walls. Much of Central Europe, but mainly in the west: most common in Ireland and western Britain.

2 Wall-rue

Asplenium ruta-muraria

Description: A small tufted fern with dark green fronds rarely more than 10cm long. The fronds are irregularly divided into diamond-shaped or fan-shaped lobes, with elongate clusters of spore capsules on the underside.
Habitat and range: Lime-rich rocks, and especially in the mortar of old walls. Throughout Europe.

3 Maidenhair Spleenwort

Asplenium trichomanes

Description: A rosette-forming fern with slender, bright green fronds up to 20cm long. Each frond has a black mid-rib and numerous oval or oblong lobes, with elongate clusters of spore capsules underneath.
Habitat and range: Rock crevices, especially on lime-rich rocks, and in the mortar of old walls. Throughout Europe.

4 Hart's-tongue Fern

Phyllitis scolopendrium

Description: A tufted fern with undivided fronds, up to 60cm long and bearing oblique rows of spore capsules underneath. The fronds are bright green at first, but darken as they get older.
Habitat and range: Damp rocks, walls, and hedge-banks. Most of Europe except the far north, but much less common on the continent than in the British Isles, where it flourishes best in the wetter climate of the west.

1

2/3

4

Grasses

Grasses have tiny flowers with no petals. They are enclosed in scaly green or brownish packets called spikelets. The latter, each containing one or more flowers, are carried in slender spikes or branching clusters. When the flowers are mature, the anthers hang from the spikelets and scatter pollen in the breeze. Several species, mostly perennials, occur as garden weeds, especially at the bases of walls and around trees and shrubs.

1

Couch Grass

Elymus repens

Description: Dull grey stems and leaves up to 120cm. The flower spike is rough to touch and consists of two rows of pointed spikelets, arranged on opposite sides of the stem with their broad sides against the main axis. This perennial species is the only really troublesome grass in the garden. It's creeping, wiry rhizomes branch rapidly through the soil and are very difficult to eliminate once the plant has become established. The rhizomes have very sharp tips and often grow right through bulbs and tubers. The grass is also called twitch.
Flowering time: June–August.
Habitat and range: Disturbed land of all kinds, including cultivated fields and gardens. Uncommon away from human habitation. All Europe.

2

Hairy Finger Grass

Digitaria sanguinalis

Description: A sprawling annual, rarely more than about 30cm, with the basal parts of the stems often lying flat on the ground. The leaves are fairly broad, usually quite hairy at the base and often tinged with purple. The flowers are carried in slender spikes that spring from the top of the stem and resemble short lengths of cord. There are usually between four and ten spikes, upright and close together at first, but gradually separating as the flowers open.
Flowering time: July–October.
Habitat and range: Waste places of all kinds, commonly invading neglected fields and gardens: particularly favours the bases of walls. Much of Europe, but most common in the south: rare in the British Isles.

3

Annual Meadow Grass

Poa annua

Description: A loosely tufted and often rather sprawling annual up to about 30cm. The bright green leaves arise from flattened shoots and are often somewhat crinkly when young. The conical flower clusters, with relatively few branches are reminiscent of tiny Christmas trees.
Flowering time: Throughout the year.
Habitat and range: Disturbed and bare ground of all kinds, commonly sprouting from crevices in paths and patios – where it lies almost flat and is largely unaffected by trampling. The grass is also common in shady lawns, and springs up in flower beds and in the vegetable patch whenever the hoe is idle for more than a few days. Throughout Europe.

1
2/3

1

Green Amaranth

Amaranthus hybridus

Description: A much-branched, deep green annual up to about 100cm. Can be mistaken for a stinging nettle (p.220) from a distance, but the leaves are oval, with smooth margins and few or no hairs. The tiny flowers are borne in dense, spiky clusters. They are green or reddish with bright yellow anthers.

Flowering time: July–October.

Habitat and range: Waste ground and neglected cultivated areas. Sometimes covers old manure heaps, from which many seeds undoubtedly reach our gardens. An American plant now well established in many parts of western Europe, especially in the south. Uncommon in the British Isles.

2

Fat Hen

Chenopodium album

Description: A stiff, branched annual, up to about 100cm, with a mealy white coating on the undersides of the leaves. The lower leaves are diamond-shaped or triangular, with irregular teeth, but the upper leaves are more or less linear and untoothed. The stems are often reddish. The flowers have no petals and are carried in pale spikes. Yellow anthers hang from the mature flowers.

Flowering time: May–October.

Habitat and range: Abundant on waste ground, manure heaps, and other disturbed areas. Can be cooked and eaten like spinach. All Europe.

Similar species: Good King Henry has untoothed triangular leaves. Common Orache has separate male and female flowers: female flowers and fruits are surrounded by two triangular bracts.

3

Black Bindweed

Bilderdykia convolvulus

Description: A slender, sprawling or climbing annual with heart-shaped or triangular leaves that are somewhat mealy on the underside. It twines clockwise. The small pink and green flowers grow in loose clusters and are followed by little black fruits with white wings.

Flowering time: June–October.

Habitat and range: Waste ground and other disturbed places. All Europe.

Similar species: Field Bindweed (p.230) is similar when not in flower, but it twines anticlockwise and its leaves are never mealy.

4

Knotgrass

Polygonum aviculare

Description: A sprawling, mat-forming annual with tough, wiry stems and more or less oval leaves, those on the flowering shoots being smaller than those on the main stems. The tiny flowers are pink or white and carried in small clusters. Dead flowers remain and enclose the fruits.

Flowering time: May–November.

Habitat and range: Bare ground, especially well-trodden paths and gateways: commonly sprouts from cracks in paving. All Europe.

1/2

3/4

1

Broad-leaved Dock

Rumex obtusifolius

Description: A deep-rooted, branching perennial, up to about 120cm, with lower leaves heart-shaped at the base. The leaf margins are slightly wavy. The flowers are small and green and arranged in whorls on the stems. As in all docks, the three inner petals enlarge after pollination, forming triangular papery valves around the 3-sided fruit and helping to carry it away on the wind. The valves in this species have spiky edges and one of them usually has a rounded swelling that can be mistaken for a seed.

Flowering time: May–October.

Habitat and range: Waste ground, roadsides, field margins, and neglected gardens. A serious agricultural weed, whose abundant seeds can survive unharmed in the soil for 50 years or more. All Europe.

Similar species: Curled Dock (below) usually has a swelling on each valve. Clustered Dock has a rather zig-zag stem. There are several other species but they all hybridise readily and identification is often difficult. Accurate identification of docks always necessitates examination of the fruits.

2

Curled Dock

Rumex crispus

Description: A branching perennial, up to about 100cm, with strongly wrinkled leaf margins. All leaves usually taper at the base. The flowers resemble those of Broad-leaved Dock but are usually carried in denser spikes. The valves surrounding the fruits are smoothly rounded and all three usually bear swellings.

Flowering time: May–October.

Habitat and range: Waste ground, roadsides, field margins, and neglected gardens: a serious agricultural weed, difficult to eradicate because of its deep roots and high seed production. All Europe.

Similar species: Broad-leaved Dock (above) has heart-shaped bases to lower leaves.

3

Greater Celandine

Chelidonium majus

Description: A much-branched perennial, up to about 100cm, with brittle stems exuding a bright orange juice when broken. The leaves are greyish green, deeply divided and more or less hairless. The short-lived flowers each have four yellow petals and two rather hairy sepals that fall as the flower opens. The seeds are carried in slender capsules.

Flowering time: April–October.

Habitat and range: Waste ground, especially in rocky and partly shaded places: rarely far from human habitation. Not uncommon on and around old garden walls. Throughout Europe. **The plant is very poisonous**.

fruit of Curled Dock. Note smooth edges of valves

fruit of Broad-leaved Dock. Note toothed edges of valves

1/2

3

1 Creeping Buttercup

Ranunculus repens

Description: A sprawling perennial, whose creeping stems root at the joints and quickly produce large colonies on inter-connected plants. Some of these colonies exceed 1.5m in diameter and are very hard to eradicate if they invade the herbaceous border. The somewhat triangular leaves are deeply lobed and the terminal lobe has a distinct stalk. The flowers have five bright yellow petals, numerous stamens, and five sepals that remain erect and form a small cup under the petals.
Flowering time: April–September.
Habitat and range: Rough grassland and disturbed ground, especially where damp: a troublesome weed on heavy soils. Most of Europe.
Similar species: The stalked terminal lobe of the leaf distinguishes this from other buttercups. Creeping Cinquefoil (below) has 5–7 toothed oval leaflets on each leaf. Silverweed (below) has pinnate leaves with paired leaflets.

2 Creeping Cinquefoil

Potentilla reptans

Description: A sprawling perennial with long, creeping, reddish stems that root at the joints and produce extensive carpets. The leaves have 5–7 toothed leaflets radiating from a central point. The solitary, bright yellow flowers have five slightly notched petals and many stamens. There are five sepals, with five more sepal-like flaps forming an epicalyx just below them.
Flowering time: May–September.
Habitat and range: Waste and bare ground, including roadside verges, paths, and neglected gardens. All Europe except the far north.
Similar species: Silverweed (below) has very different leaves. Creeping Buttercup (above) has no epicalyx.

3 Silverweed

Potentilla anserina

Description: A sprawling perennial, whose creeping stems root at the joints. The leaves are silvery underneath and each has 7–12 pairs of oval, toothed leaflets. Bright yellow flowers are borne singly on long reddish stalks. Each flower has five petals, which are oval or rounded and not notched as in most other cinquefoils, five sepals, and an epicalyx below them (see above).
Flowering time: May–August.
Habitat and range: Waste ground, roadside verges, neglected gardens, and other disturbed land, especially in damp, trampled areas. Most of Europe except the far north and south.
Similar species: The divided silvery leaves distinguish this from all other yellow-flowered weeds.

Creeping Buttercup leaf. Note the terminal leaflet has a distinct stalk – this feature is diagnostic

1

2

3

1

Yellow Oxalis

Oxalis corniculata

Description: A sprawling perennial with stems rooting at the joints. The leaves are green or purplish brown, with 3 heart-shaped leaflets. The yellow flowers are followed by dagger-shaped fruits on strongly bent stalks. Also called Yellow Sorrel.

Flowering time: May–October.

Habitat and range: Mainly on bare ground. The purple form is especially common on paths and patios and also in flower pots. Much of Europe, but rare in north, where it occurs mainly in and around gardens.

Similar species: O. stricta does not root at the joints. Upright Yellow Oxalis does not root at joints and has straight fruits stalks.

2

Common Fumitory

Fumaria officinalis

Description: A short, weak-stemmed annual with much-divided, greyish green, hairless leaves. The tubular pink flowers, up to 9mm long, become deep purple towards the mouth and are carried in fairly dense spikes. The spikes, sometimes with 100 or more flowers, are longer than their stalks.

Flowering time: March–October.

Habitat and range: Disturbed ground of all kinds: an abundant weed of arable land and gardens. Throughout Europe.

Similar species: There are many very similar species, usually with shorter flowering spikes but longer individual flowers.

3

Yellow Corydalis

Corydalis lutea

Description: A tuft-forming perennial, up to about 30cm, with deeply divided, greyish green leaves that are easily mistaken for fern fronds. The bright yellow, drooping, tubular flowers are up to 20mm long and carried in lax spikes just above the foliage.

Flowering time: April–November.

Habitat and range: Shaded, rocky places, especially on limestone: common on many old walls. A native of the Alps, now well established throughout Southern and Central Europe.

4

Procumbent Pearlwort

Sagina procumbens

Description: A mat-forming perennial with narrow, shiny leaves and creeping, rooting stems spreading from a central rosette. Petals are often absent from the tiny green flowers and the sepals spread flat around the ripe fruit capsules. Flowers only on side branches.

Flowering time: May–September.

Habitat and range: Bare ground, especially in damp, shady spots: common on paths and patios. Throughout Europe.

Similar species: Annual Pearlwort does not form obvious rosettes and stems do not root. Main stems end in flowers.

1

2

3/4

1

Common Chickweed

Stellaria media

Description: A weak, often sprawling annual with paired bright green, oval leaves, but otherwise rather variable. A single line of hairs runs down the stem between each joint. The star-like white flowers, carried in small clusters among the upper leaves, usually have five deeply-cleft petals, but petals are sometimes missing.
Flowering time: All year.
Habitat and range: Disturbed ground of all kinds, including manure heaps and shingle beaches: a very common, fast-growing garden weed that often quickly smothers freshly-dug ground. All Europe. The plant can be cooked and eaten as a tasty alternative to spinach.

2

White Campion

Silene alba

Description: A much branched, rather weak-stemmed annual or short-lived perennial, up to 100cm and clothed with soft, sticky hairs. Its paired leaves are more or less oval and sharply pointed. The flowers, up to 30mm across, are carried in loosely branched clusters, with male and female flowers on separate plants. Female flowers are followed by swollen seed capsules reminiscent of tiny onions.
Flowering time: April–October.
Habitat and range: Hedgerows, roadsides, waste land, and neglected gardens. Throughout Europe except the far north.

3

Shepherd's Purse

Capsella bursa-pastorum

Description: A rosette-forming annual with deeply-toothed basal leaves and a simple or branched flower-head up to 40cm. The small white flowers have four petals and are followed by the heart-shaped seed capsules that give the plant its name.
Flowering time: Throughout the year.
Habitat and range: Waste ground and other bare places, including field margins and neglected gardens, especially in trampled areas. All Europe.

4

Common Whitlow Grass

Erophila verna

Description: A slightly hairy, rosette-forming annual, with lightly toothed basal leaves. The flowering stems are leafless and up to about 20cm. The small white or pale pink flowers have four deeply-cleft petals and are followed by flattened, oval seed capsules on long stalks.
Flowering time: March–May.
Habitat and range: Mostly on bare, stony or sandy ground, but not uncommon on old garden walls and rockery stones: also invades bare patches on lawns. All Europe except the far north.

1/3

2

4

1

Hairy Bittercress

Cardamine hirsuta

Description: A rosette-forming annual with flowering stems up to about 30cm. Each leaf has 3–7 pairs of rounded or diamond-shaped leaflets and a more or less kidney-shaped terminal leaflet. All are slightly hairy. The small white flowers have four petals and are followed by long, slender seed capsules.
Flowering time: February–November.
Habitat and range: Bare ground, especially in stony places: an abundant and fast-growing garden weed, spreading rapidly by shooting its seeds all over the place whenever the ripe capsules are touched. All Europe.

2

Annual Wall Rocket

Diplotaxis muralis

Description: A hairless, rosette-forming annual or biennial with deeply toothed or lobed leaves. The flowering stems, which may have a few bristles at the base, reach a height of about 60cm and rarely have any leaves or bracts. The 4-petalled yellow flowers are followed by slender fruits that are much longer than their stalks. The plant is also called wall cress and stinkweed – because of its pungent, fox-like smell.
Flowering time: May–September.
Habitat and range: Waste ground and rocky places, including old walls: also a weed of arable land and neglected gardens. A native of southern and central Europe, naturalised in the British Isles and parts of the north.
Similar species: Perennial Wall Rocket has more deeply divided leaves not in a rosette, and fruits are shorter than their stalks.

3

Biting Stonecrop

Sedum acre

Description: A hairless, mat-forming perennial with fleshy, cylindrical leaves clasping the stems. The leaves and shoots, named for their peppery taste, often turn red in autumn. The bright yellow flowers are carried in small clusters.
Flowering time: May–July.
Habitat and range: Sandy and stony places: often grown on rockeries, from which it spreads to paths, old walls, and roof-tops. All Europe.

4

Scarlet Pimpernel

Anagallis arvensis

Description: A sprawling, almost hairless annual with shiny, oval, unstalked leaves that normally grow in opposite pairs. The leaves have black dots on the underside. The star-like, solitary flowers are usually scarlet, but occasionally blue or purple. They open only in sunshine, giving the plant its alternative name of poor man's weatherglass.
Flowering time: May–October.
Habitat and range: Bare and disturbed ground, including sand dunes, field margins, and gardens: usually on well-drained soils. All Europe.

1

2/3

4

1

Common or Stinging Nettle

Urtica dioica

Description: A square-stemmed perennial, up to 150cm, with paired, strongly-toothed leaves. The plant is clothed with hollow, glassy hairs, whose tips break off when touched to leave sharp spikes that penetrate the skin and inject an irritating mixture of histamine and formic acid. Drooping clusters of male and female flowers are borne on separate plants. Creeping stems and wide-ranging yellow roots enable the plants to form large clumps.
Flowering time: May–September.
Habitat and range: Disturbed ground, especially where the soil has been enriched by burning or by the dumping of rubbish: most common near human habitation. Throughout Europe. The leaves support the caterpillars of several butterflies and can be eaten like spinach.
Similar species: Annual Nettle (below) has relatively long leaf stalks.

2

Annual or Small Nettle

Urtica urens

Description: An annual, up to 60cm. The leaves are shorter than those of the Common Nettle, although they have longer stalks – especially on the lower part of the plant where the leaf-blades are noticeably shorter than their stalks. The stings are less painful than those of the Common Nettle. Male and female flowers are borne on the same plant.
Flowering time: May–September.
Habitat and range: Gardens and other disturbed places, mainly on well-drained and relatively dry soils. Throughout Europe.
Similar species: Stinging Nettle (above) has larger leaves.

3

Ivy-leaved Toadflax

Cymbalaria muralis

Description: A sprawling, hairless perennial with lobed, long-stalked, often purplish leaves. The mauve and yellow flowers resemble tiny faces.
Flowering time: April–October.
Habitat and range: Rocky places, especially in shady spots: common on old walls, from which it spreads to gravel paths and paved areas. A native of southern Europe, now established in most parts except the far north.

4

Red Valerian

Centranthus ruber

Description: A greyish green, hairless perennial with slightly fleshy and waxy leaves. The upper leaves clasp the stem and are lightly toothed at the base. The red or pink flowers are borne in dense heads and are followed by feathery fruits that are scattered by the slightest breeze.
Flowering time: May–September.
Habitat and range: Rocky places, including old walls and roadside banks. A Mediterranean plant, now well established in many other parts of Europe. A good butterfly attractant, it is often planted in gardens, but it is an invasive plant and not easy to control.

1/2

3/4

1

Petty Spurge

Euphorbia peplus

Description: A hairless annual, up to 30cm, with bright green, round or oval leaves. The stem exudes a milky white juice when broken. The tiny flowers have no petals and several are carried in each cup-shaped cluster of yellowish, leaf-like bracts.
Flowering time: May–November.
Habitat and range: Disturbed soil and waste ground: abundant in gardens, often springing up immediately after digging. Most of Europe. **Poisonous**.

2

Caper Spurge

Euphorbia lathyrus

Description: A hairless, bluish green biennial, up to 150cm. Exudes a thick white juice when broken. Unbranched in its first year, when the leaves clustering vertically around the sturdy stem give it a likeness to a miniature poplar, but becomes much-branched in its second year. The little green flowers are surrounded by leaf-like bracts at the tips of the branches and are followed by rounded fruits that explode when ripe and fire their seeds everywhere with great force. The fruits resemble the capers used in cookery, but **they are very poisonous**.
Flowering time: May–August.
Habitat and range: Disturbed and waste ground: sometimes planted in gardens because it is reputed to deter moles – and seems to work in some places. A native of southern and central Europe, now widely naturalised.

3

Common Cleavers

Galium aparine

Description: A weak-stemmed annual, scrambling over other plants with the aid of minute hooks on its stems and its whorled leaves. Tiny white flowers are born in small clusters and are followed by rounded, olive-green fruits that are covered with hooked bristles. The latter catch in almost anything that touches them and help to carry the fruits away.
Flowering time: May–August.
Habitat and range: Hedgerows, waste ground, and most disturbed places. Can be a severe nuisance in shrubberies and herbaceous borders if allowed to become established. All Europe except the far north.

4

Common Mallow

Malva sylvestris

Description: A rather variable, hairy, deep-rooted biennial or short-lived perennial, up to 150cm, but usually shorter and often sprawling. The lower leaves are kidney-shaped or rounded with several shallow lobes, while the upper ones are deeply cut. The deeply-notched petals are pink or light purple with darker stripes, and are well separated from each other.
Flowering time: May–October.
Habitat and range: Waste ground, roadsides, and neglected gardens. All Europe except the far north.

1

2

3/4

1

Herb Robert

Geranium robertianum

Description: A much-branched, hairy annual or biennial, up to 50cm but often sprawling. Stems usually tinged with red: leaves deeply divided, with 3 or 5 lobes, and often quite red. The flowers are borne in pairs and the petals are bluntly rounded at the tips, with no more than a very faint notch. They are usually pink, but white forms occur here and there. The whole plant has a pungent smell, not unlike that of a fox.
Flowering time: May–September.
Habitat and range: Woods and other shady places, often among rocks: in the garden it is most frequent in shrubberies and hedge bottoms and at the base of old walls. All Europe except the far north.

2

Shining Crane's-bill

Geranium lucidum

Description: A much branched, low-growing annual with shiny leaves often strongly tinged with red. The leaves are rounded in outline but divided into 5 more or less oval lobes, with the incisions reaching no more than about half way into the blade. The petals are pink and oval, without a terminal notch. As in all crane's-bills, the fruits have long 'beaks'.
Flowering time: April–September.
Habitat and range: Hedgebanks, old walls, and other rocky places, usually in the shade but sometimes spreading to the open garden: mainly on lime-rich soils. All Europe except the far north.

3

Cut-leaved Crane's-bill

Geranium dissectum

Description: A sprawling or upright annual or biennial. Very hairy, with the upper leaves divided into several finger-like lobes: the lobes of the lower leaves often fork again. The petals are deep pink and distinctly notched. Fruits and the surrounding sepals are very hairy.
Flowering time: April–September.
Habitat and range: Waste ground, dry grassland, and as a weed of cultivated land – especially the bare, trodden areas around the edges. All Europe except the far north.

4

Enchanter's Nightshade

Circaea lutetiana

Description: A slender, slightly hairy perennial up to 70cm, with more or less oval, opposite leaves. The flowers, borne in elongated clusters well above the leaves, are white or pale pink and have only 2 petals, although each is deeply divided to give the impression of 4 petals. The plant spreads rapidly by means of creeping stems at or just below ground level and can be a real nuisance in the garden.
Flowering time: June–September.
Habitat and range: Woods and other shady places, including garden shrubberies and neglected borders. Most of Europe. A member of the willowherb family, the plant is unrelated to the true nightshades.

1/2

3/4

1 Rosebay Willowherb

Epilobium angustifolium

Description: A slender perennial, up to 250cm, with narrow leaves carried alternately nearly all the way up the stem. The flowers are bright pink, often tinged with purple, and are borne in long, tapering heads. The four petals are only slightly notched and the stamens and stigmas protrude well beyond them. Flowers are followed by slender capsules that split to released clouds of feathery seeds. Quick-growing horizontal roots enable the plant to form large clumps in a short time.

Flowering time: June–September.

Habitat and range: Almost anywhere, but especially on disturbed ground: often called fireweed because of its liking for burnt ground. Frequently establishes itself on old walls, where it can damage mortar with its spreading roots. All Europe.

2 Broad-leaved Willowherb

Epilobium montanum

Description: A slender, slightly hairy perennial up to 60cm. The lower leaves are in opposite pairs, but the upper ones are borne alternately. Rounded at the base, they have very short stalks and are usually lightly toothed. The flowers are more or less upright, with 4 pale pink petals that are conspicuously notched. The stigma is clearly 4-lobed. The flowers are followed by long, slender capsules that split to release feathery seeds. Fleshy pink or white stems spread from the base at or just below ground level, mainly in autumn.

Flowering time: May–August.

Habitat and range: Almost anywhere, but especially on disturbed ground: common on old walls. All Europe.

Similar species: Pale Willowherb has white flowers, often streaked with pink, and its leaves become narrow at the base. American Willowherb has ridged stems and the stigma is club-shaped.

3 Ground Elder

Aegopodium podagraria

Description: A hairless perennial, up to 100cm (but usually much less) and carpeting the ground with leaves sprouting from its tangled, fast-growing rhizomes. The leaves have somewhat fleshy stalks and have 3 main divisions, each of which may be further divided into 2 or 3 toothed leaflets. The small white flowers are packed into domed heads (umbels) up to 6cm across.

Flowering time: June–August.

Habitat and range: Shady places almost anywhere, but in the British Isles it is most common in and around human settlements, where it was once cultivated: although rather strong, it can be eaten like spinach. The plant is now a formidable garden weed, very difficult to eradicate because even the smallest piece of rhizome left in the ground can quickly grow into a new plant. Most of Europe.

1

2

3

1

Giant Hogweed

Heracleum mantegazzianum

Description: A stout biennial with stems up to up to 10cm in diameter and 500cm high. The leaves, up to 100cm long, are divided into several jagged leaflets. The white flowers are carried in umbrella-like heads (umbels) up to 50cm across. The sap can cause serious skin blisters.

Flowering time: June–September.

Habitat and range: A native of south-west Asia, it originally came to Europe as a garden plant, but its seeds are easily carried on the wind and it is now established in many habitats, including river banks, waste ground, and roadsides. It probably springs up more commonly on allotments than in private gardens. Most of Europe.

2

Cow Parsley

Anthriscus sylvestris

Description: A much branched, downy biennial or perennial with hollow, furrowed stems up to 100cm high and dull green leaves that are finely divided and often mistaken for fern fronds. The flowers are usually white, but occasionally pale pink, and are carried in umbels up to 6cm across. Their intricacy led to the alternative name of Queen Anne's Lace. The plant emits a spicy odour when crushed.

Flowering time: April–June – earlier than most similar plants.

Habitat and range: Waste ground and rough grassland: abundant in hedgerows and roadside verges, from which it readily invades gardens and orchards. All Europe except the far north.

Similar species: There are many superficially similar species but most are either hairless or have solid stems. The only one commonly found in the garden is Fool's Parsley (below).

3

Fool's Parsley

Aethusa cynapium

Description: A bright green, hairless annual with a finely ridged hollow stem up to 100cm. The leaves, more or less triangular in outline, are divided into numerous oval leaflets. Umbels of white flowers are often flat and long green 'whiskers' hang down from the individual flower clusters.

Flowering time: June–October.

Habitat and range: Waste ground, field margins, and gardens. All Europe except the far north. The plant is **very poisonous** and care must be taken to distinguish it from cultivated herbs such as coriander and parsley, especially the flat-leaved variety that is now very popular. It is easy when the plants are in flower, but smell is the best guide at other times: Fool's Parsley does not smell.

Fool's Parsley flower. Note the 'whiskers' hanging down from the flower clusters

1

2

3

1

Field Bindweed

Convolvulus arvensis

Description: A climbing or sprawling perennial with stems up to 200cm long. It twines anticlockwise up any suitable support. The leaves, shaped like arrowheads, are often greyish or bluish green. The funnel-shaped flowers are white or pink and each opens for just a few hours.
Flowering time: May–September.
Habitat and range: Waste ground, rough grassland, and cultivated areas, often climbing cereal stalks at the edges of fields: a real nuisance if it gets a hold in the herbaceous garden or the shrubbery, for its very deep, fleshy rhizomes make it difficult to eradicate.
Similar species: The unrelated Black Bindweed (see p.208) has similar leaves but its flowers are very different and it twines clockwise.

2

Large Bindweed

Calystegia silvatica

Description: A climbing perennial often reaching heights of 300cm on fences and hedges. The stems twine in an anticlockwise direction. Leaves are large and shiny and shaped like arrowheads. The funnel-shaped white flowers, up to 9cm across and sometimes lightly striped with pink on the outside, are very attractive. The base of each flower is enclosed in two overlapping leaf-like bracts.
Flowering time: June–September.
Habitat and range: Waste ground and hedgerows, from where it invades gardens and may smother shrubberies and herbaceous borders and fences. A native of southern Europe, introduced to the British Isles as a garden plant and now widely distributed in the southern half of England.
Similar species: Hedge Bindweed (*C. sepium*) is very similar but the flowers are only about 5cm across and the bracts at the base hardly overlap (see illustration below). It sprawls over hedges and fences with Large Bindweed, but inhabits a wider range of places, including river banks, marshes, and woodland margins. The two species probably hybridise.

3

Ground Ivy

Glechoma hederacea

Description: A creeping, mat-forming perennial, rooting at the joints and producing erect flowering stems up to 30cm high. It is usually clothed with soft hair and is strongly aromatic. The more or less heart-shaped leaves are strongly toothed and have long stalks. The flowers, borne in small clusters in the leaf axils, are blue or violet with purple spots on the lower lip.
Flowering time: March–July.
Habitat and range: Light woodland, hedgebanks, waste ground, and neglected gardens: often invades damp, unkempt lawns. All Europe.

flower of Large Bindweed. Note overlapping bracts

flower of Hedge Bindweed. Note bracts hardly overlap

1

2

3

1

Self-heal

Prunella vulgaris

Description: A creeping, mat-forming perennial producing erect flowering stems up to 30cm high. The leaves are oval or diamond-shaped, with smooth or lightly toothed edges. The plant is only slightly hairy and is not aromatic. The flowers are usually purple or violet and are borne in roughly rectangular heads at the tops of the stems.

Flowering time: May–November.

Habitat and range: Grassy places of all kinds, including orchards and garden lawns. All Europe. The plant was once used for dressing wounds.

2

Henbit Deadnettle

Lamium amplexicaule

Description: A hairy annual up to 25cm, with few branches. The leaves range from round to almost triangular, with blunt teeth: upper ones are unstalked and completely surround the stem. Pink flowers have long, slender tubes and grow in loose whorls among the upper leaves.

Flowering time: March–November.

Habitat and range: Waste ground, field margins, and gardens – especially bare patches in the vegetable plot. Mainly on light soils. All Europe.

3

Red Deadnettle

Lamium purpureum

Description: A downy, much branched, and strongly aromatic annual up to 45cm. The leaves, which are often tinged with purple, are all stalked. They are more or less heart-shaped, with rounded teeth and a wrinkly surface. The flowers are some shade of pink, with a hood and a deeply-divided lower lip with only 2 obvious lobes. They are borne in dense whorls at the tops of the stems.

Flowering time: All year, but flowers are most abundant in spring.

Habitat and range: Bare ground in waste places and on cultivated land: a common garden weed, quickly invading bare patches.

Similar species: Henbit Deadnettle (above) has upper leaves clasping the stem. Spotted Deadnettle is a commonly cultivated perennial, usually with white spots on the leaves.

4

White Deadnettle

Lamium album

Description: A vigorous, hairy, patch-forming perennial up to 60cm, with creeping stems on or just under the ground. The leaves are of an elongated heart-shape and strongly toothed, with the flowers in whorls in the axils. The flower's upper lip or hood is hairy and the lower lip has two prominent, rounded lobes. The plant is slightly aromatic.

Flowering time: March–November.

Habitat and range: Waste ground, hedgebanks and verges, and neglected gardens. All Europe except the far north.

Similar species: Leaves resemble those of Stinging Nettle (p.220), but do not sting.

1/2
3
4

1 Black Nightshade

Solanum nigrum

Description: A hairless or slightly downy annual, up to 60cm but often rather straggly. The leaves range from oval to more or less triangular, with smooth or lightly indented margins. The starlike white flowers have yellow anthers in the middle and are borne in small clusters. They are followed by **poisonous** green berries that turn black as they ripen.
Flowering time: June–October.
Habitat and range: Waste ground and other disturbed places: a common field and garden weed. All Europe except the far north. The berries can ripen and release seeds even after the weeds have been pulled up, so the plants should be completely destroyed.
Similar species: Hairy nightshade is much hairier, with orange or brown berries.

2 Thornapple

Datura stramonium

Description: A usually hairless annual up to 150cm, with forking branches and large, jagged leaves. The white, trumpet-shaped flowers, up to 8cm across, are born singly and each one last for just a few hours. They are followed by the very spiny, egg-shaped seed capsules that give the plant its name.
Flowering time: June–October.
Habitat and range: Waste ground and other bare places, mostly in and around human settlements. A native of south-east Europe, now well established in most parts except Ireland and the far north, but usually sporadic in appearance in the cooler regions. The plant is **very poisonous**. It was once burned in smithies, where the smoke was said to calm the horses about to be shod, and this may well have been a source of seed for the surrounding fields and gardens. The seeds can lie dormant in the soil for many years.

3 Germander Speedwell

Veronica chamaedrys

Description: A sprawling and rather hairy perennial up to 50cm high. The stems are often reddish brown and have two distinct lines of hairs, on opposite sides. The leaves are more or less oval, with coarse teeth on the edges, They have no more than short stalks and frequently clasp the stem. The bright blue flowers, each with a white eye, are borne in elongated clusters springing from the bases of the upper leaves. As in all speedwells, individual flowers do not last for more than a few hours.
Flowering time: March–July.
Habitat and range: Waste ground, hedgebanks, and other rough places, including neglected gardens. All Europe except the far north.
Similar species: Apart from Wall Speedwell, other garden speedwells bear their flowers singly (see p.236).

1

2

3

1

Common Field Speedwell

Veronica persica

Description: A sprawling, hairy annual with more or less triangular, coarsely-toothed leaves on short stalks. The bright blue flowers are borne singly on long stalks in the axils of the leaves. The lowest petal is smaller than the rest and usually white.
Flowering time: All year.
Habitat and range: Mostly a weed on cultivated ground, where it is usually the commonest of all speedwells: often covers gardens and allotments in winter before the digging begins. A native of south-west Asia, now well-established throughout Europe except the far north.
Similar species: Grey Field Speedwell has greyer leaves and flowers are completely blue. Green Field Speedwell has the upper petal bright blue and the other 3 very pale. See also Ivy-leaved Speedwell (below).

2

Ivy-leaved Speedwell

Veronica hederifolia

Description: A sprawling, downy annual whose leaves have 3–7 lobes and resemble small ivy leaves. The flowers are lilac or pale blue and are carried singly on fairly short stalks – rarely longer than the leaves.
Flowering time: March–September.
Habitat and range: Waste ground and bare soil, especially on cultivated land. A very common garden weed, seeding profusely and quickly covering low-growing plants in the herbaceous border and rockery if unchecked. Commonly transported with container-grown plants from garden centres.

3

Slender Speedwell

Veronica filiformis

Description: A creeping, downy, mat-forming perennial with stems rooting at the joints. The leaves are roughly kidney shaped, with rounded teeth and short stalks. The flowers are borne singly on slender stalks that are much longer than the leaves. The upper petal is violet blue but the others are pale and the lowest one is almost white.
Flowering time: April–August.
Habitat and range: Grassy places: abundant on many lawns. A native of western Asia, brought to Europe as a rockery plant and now established throughout southern and central Europe. It rarely sets seed, but spreads rapidly by fragmentation of its creeping stems.

4

Wall Speedwell

Veronica arvensis

Description: A tufted, downy annual up to 25cm, with coarsely toothed triangular or oval leaves. The bright blue flowers cluster tightly among leaf-like bracts around the upper parts of the stems.
Flowering time: March–October.
Habitat and range: Open areas, including well-drained lawns, old walls, and gravel paths: usually in dry places. All Europe.

1

2/3

4

1 Ribwort Plantain

Plantago lanceolata

Description: A rosette-forming perennial, with strap-shaped leaves up to 30cm long. The leaves are hairless or slightly downy and they narrow gradually towards the base. Almost flat in trampled turf, they are more or less upright in the longer grass of orchards and hedge bottoms. The tiny flowers are carried in dense, dark spikes at the top of leafless stems up to 50cm high. The flowers open successively from the bottom upwards, so a ring of dangling cream stamens seems to move gradually up the spike.
Flowering time: April–October.
Habitat and range: Waste ground and grassy places, including many lawns: especially common on roadside verges. Throughout Europe.

2 Greater Plantain

Plantago major

Description: A rosette-forming perennial with broad, hairless or slightly downy leaf blades that are sharply demarcated from their stalks. The leaves are flat on trampled areas, but more or less upright in longer grass. The tiny flowers are carried in long, slender spikes that give the plant its alternative name of rat-tailed plantain. The flowering stems are up to 50cm high. The lowest flowers open first and a ring of purplish stamens gradually moves up the spike, as in Ribwort Plantain (above).
Flowering time: May–October.
Habitat and range: Waste ground and grassy places, including lawns and playing fields. Very tolerant of trampling, it commonly springs up on paths, drives, and pavements, and also on bare patches in the vegetable garden. Throughout Europe.
Similar species: Hoary Plantain (below) is very downy.

3 Hoary Plantain

Plantago media

Description: A rosette-forming perennial clothed with silvery down and having a distinctly grey appearance. The leaf blades are oval and narrow gradually into a short stalk. They nearly always lie flat on the ground. The flowering stems are up to 30cm high, with dark flower spikes like those of Ribwort Plantain (above). The stamens are pale lilac and, unlike those of other plantains, the flowers have a delicate scent.
Flowering time: May–September.
Habitat and range: Waste ground and grassy places, usually where the grass is short: not uncommon on lawns, but only on lime-rich soils.
Similar species: Greater Plantain (above) is much greener.

1/2

3

1

Groundsel

Senecio vulgaris

Description: A hairless or slightly downy, weak-stemmed annual, up to 50cm high. The bright green leaves are deeply lobed or divided and the upper ones clasp the stem. Minute yellow flowers or florets are packed into tight heads, each about 5mm in diameter and surrounded by black-tipped bracts, and the flowerheads are grouped into loose clusters at the tops of the stems. The flowers are followed by fluffy white fruits. The leaves are often infested by a bright orange rust fungus.

Flowering time: All year.

Habitat and range: Waste ground and other disturbed areas: a very common garden weed, spreading rapidly by its feathery fruits. All Europe.

2

Oxford Ragwort

Senecio squalidus

Description: A much-branched annual or short-lived perennial up to 30cm high. The deeply-lobed leaves are hairless or slightly hairy: the upper ones are stalkless and clasp the stem and the lower ones have winged stalks. Numerous flowerheads are carried in domed or flat-topped clusters. Each flowerhead usually has 13 bright yellow ray florets surrounding a larger number of tiny tubular florets, with black-tipped bracts underneath them. Fluffy white fruits follow the flowers and carry their seeds over long distances.

Flowering time: April–December.

Habitat and range: Waste ground, railway lines, and central reservations of many main roads: often springs up at the bottoms of fences and walls in towns, and also grows on old walls. A native of Sicily, now widely distributed in the British Isles and many other parts of central Europe.

3

Pineapple Mayweed

Chamomilla suaveolens

Description: A much-branched, hairless annual up to about 35cm high. The leaves are deeply divided into slender segments and have a feathery appearance. Numerous yellowish green flowers (florets) are packed into domed or conical heads, up to 8mm across, at the tips of the branches. The whole plant smells strongly of pineapples when crushed.

Flowering time: April–November.

Habitat and range: Waste ground and most other disturbed areas, including field margins and roadsides: tolerant of trampling, it often flourishes in cracks in paths and patios. A native of eastern Asia, it is now well-established in most parts of Europe.

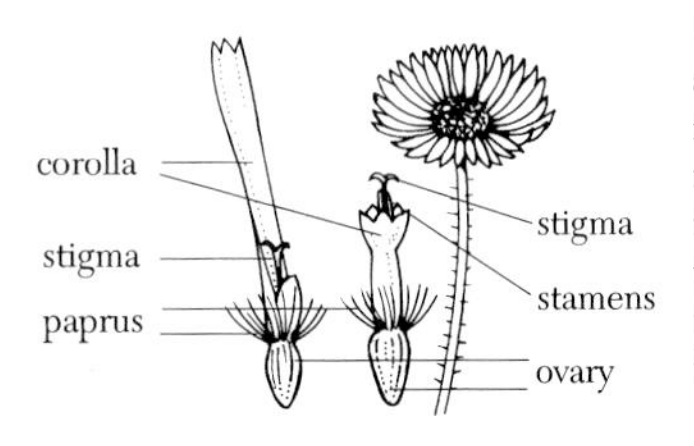

Members of the daisy family are called composites. Their flowerheads consist of numerous tiny flowers called florets. There are two kinds: tubular florets (right) and ray florets (left) with a tongue-like flap on one side

1

2

3

1

Nipplewort

Lapsana communis

Description: A rather slender and often much-branched, slightly hairy annual up to about 100cm. Lower parts are usually hairy; upper stems and leaves are hairless. Lower leaves have long stalks and are deeply lobed, with a very large, toothed terminal lobe; upper leaves oval or diamond-shaped, usually with blunt teeth and often unstalked. Stems do not contain milky latex. Flowerheads, with up to 15 pale yellow florets, are borne in open clusters on slender stalks. Unopened heads resemble small nipples.
Flowering time: June–October; closes in dull weather.
Habitat and range: Disturbed ground of all kinds: a very common garden weed, especially in the vegetable plot, where regular digging enables it to colonise very quickly. Throughout Europe.

2

Creeping Thistle

Cirsium arvense

Description: A spiny-leaved perennial, usually forming distinct patches up to about 100cm high. The stems are always rounded, without wings or spines. The leaves, hairless on the upper surface but often downy underneath, are usually deeply lobed and very spiny. The flowerheads, each containing numerous pale purple or lilac florets, are usually borne in small clusters. A single plant can rapidly develop into a thick clump by means of its creeping roots, small pieces of which can also grow into new plants.
Flowering time: June–September.
Habitat and range: Disturbed ground of all kinds: a serious weed of arable fields and allotments. Throughout Europe.

3

Smooth Sowthistle

Sonchus oleraceus

Description: Greyish, hairless annual up to 150cm, with brittle stems and soft, spiny teeth on the leaf margins. Lower leaves are usually deeply lobed, with a triangular terminal lobe that is much wider than the rest. Upper leaves are often undivided and clasp the stem with their pointed bases. The flowerheads contain many golden yellow ray florets grouped into small clusters. These do not all open at the same time and are followed by fluffy white fruits. As with all sowthistles, it exudes a white latex when broken.
Flowering time: May–November.
Habitat and range: Waste ground and other places with bare soil: a common weed of gardens and allotments, frequently sprouting from pavement cracks and old walls in rural areas. All Europe except far north.
Similar species: Prickly Sowthistle has much harder spines and the leaf bases bend back to clasp the stem with rounded lobes.

leaf of Prickly Sowthistle. Note the leaf-base curls back

leaf of Smooth Sowthistle. Note the leaf-base sticks forward

1

2

3

1

Mugwort

Artemisia vulgaris

Description: A slightly downy, aromatic perennial up to 120cm high and arising from a somewhat woody base. The stems are often reddish brown or purple. The deeply-divided, dark green leaves are hairless above but clothed with felt-like white hair underneath. The pale, oval flowerheads, containing tiny brown or yellowish florets, are carried in large, branching clusters.

Flowering time: May–September.

Habitat and range: Waste ground and other disturbed places, including riverbanks and roadsides: often springs up in hedgerows and under old garden walls and quickly invades neglected gardens and allotments.

2

Feverfew

Tanacetum parthenium

Description: A much branched, slightly hairy biennial or short-lived perennial up to about 75cm high. The whole plant is strongly aromatic and often has a slightly sticky feel. Mature plants become very woody at the base. The leaves are usually yellowish green and are deeply divided. The daisy-like flowerheads have fairly long stalks and are carried in domed clusters. Each head is 10–25mm across, with about a dozen ray florets surrounding a mass of yellow disc florets.

Flowering time: May–October.

Habitat and range: Waste ground and other disturbed places, often growing on old garden walls. Winged fruits enable it to colonise gardens very easily. A native of south-east Europe, now established in many parts of southern and central Europe. The plant's name is said to reflect its use in combating fever. Eaten in small quantities, the leaves are also believed to alleviate migraine.

3

Yarrow

Achillea millefolium

Description: A perennial with woolly stems up to 75cm high. Horizontal stems at or just below ground level grow rapidly and enable the plant to form extensive patches. The leaves are deeply divided into feathery lobes and sometimes mistaken for fern fronds. The flowerheads resemble small daisies packed into dense domed or flat-topped clusters. Each head usually has 5 white or occasionally pink ray florets surrounding a bunch of white or cream disc florets. The whole plant has a strong aromatic scent.

Flowering time: June–October.

Habitat and range: Waste ground and grassy places. It is common on lawns, where it forms extensive carpets although regular mowing prevents its flowering. Throughout Europe.

1/2

3

Lichens

Lichens are very unusual organisms. Each one consists of an intimate association between a fungus and an alga, with both partners benefiting from the relationship. The algae involved are microscopic green plants similar to those that form bright green stains on damp tree trunks or turn pond water green in summer. These algae can exist on their own, making food by photosynthesis, but the fungi depend entirely on the algae for food. For this reason, lichens are now regarded simply as fungi with an unusual way of getting food. They grow on a wide range of substrates, including bricks, concrete, tree trunks, and even asbestos and roofing felt. Despite their make-up, lichens do not resemble either fungi or algae. They are mostly tough, dry, slow-growing organisms. The lichen body is called the thallus and occurs in three main forms. Crustose lichens form thin crusts; foliose lichens form patches of radiating lobes, and fruticose lichens are like miniature bushes. Powdery granules produced on the surface contain both fungus and alga and are blown away to grow directly into new lichens. The fungus partner can also scatter spores from certain, often colourful parts of the lichen surface, but the germinating spores soon die unless they link up with the right kinds of algae. Lichens are incredibly hardy organisms, surviving extreme heat and cold as well as drought, but few species can tolerate air pollution. A lens is essential for identifying many of the species.

1

Caloplaca heppiana

Description: A golden yellow crustose lichen, forming irregular patches with lobed margins that are a little paler than the rest of the thallus. The tips of the lobes are slightly swollen. Orange-brown spore-bearing discs called apothecia develop in the middle of the patches.
Habitat and range: Common on limestone rocks and on old walls and concrete surfaces, including many old gravestones. Most of Europe.
Similar species: There are many superficially similar lichens. *C. aurantiana* is paler and the margins are not swollen. *Xanthoria* species (below) are similarly coloured but are foliose lichens composed of branching lobes.

2

Xanthoria parietina

Description: A foliose species with fairly loose orange-yellow lobes, often but not always forming circular patches. Older specimens often die off in the middle to leave a ring or a cluster of irregular patches. The spore-bearing discs are the same colour as the rest of the thallus or slightly darker.
Habitat and range: More tolerant of air pollution than most other similar lichens, it grows on rocks, walls, tree trunks, and asbestos roofs. All Europe.

3

Xanthoria aureola

Description: A foliose lichen, very similar to *X. parietina* except that the surface is clothed with short rods – visible only with a good lens.
Habitat and range: Walls and rocks. Throughout Europe.

1

2

3

1

Ochrolechia parella

Description: A crustose lichen, forming a rough greyish brown crust with a conspicuous white margin. Starting off as a more or less circular patch, it expands in whatever direction it can and may produce a large and very irregular crust as much as 15cm across. The spore-producing discs are pinkish brown with thick, wavy white margins.
Habitat and range: Rocks and walls of all kinds: not uncommon in towns. Throughout Europe.
Similar species: Lecanora calcarea and *L. campestris* are greyer and their crusts have conspicuous cracks, like miniature crazy paving. Their spore-bearing discs are much darker.

2

Lecanora muralis

Description: Although this is a foliose lichen, the leaf-like lobes are noticeable only at the margins of the thallus, where they are firmly attached to the substrate. The thallus is often markedly circular and is brownish grey with a paler margin. Numerous brown spore-bearing discs with irregular pale borders are crowded into the central region.
Habitat and range: Rocks and walls, especially those with plenty of lime-rich mortar. Also on gravestones, asbestos roofs, and concrete paths – including pavements if they are not too heavily used. Fairly tolerant of pollution, it is not uncommon in towns. Throughout Europe.

3

Lecanora conizaeoides

Description: A crustose lichen forming extensive, much-cracked, greenish-grey crusts. The spore-bearing discs are pale brown or yellowish green with thick, 'pimply' margins.
Habitat and range: Grows on tree trunks and branches, fences, and walls. More tolerant of pollution than almost any other lichen, it can be found right in the centre of towns, where it is often the only lichen growing on wood. It is quite rare in unpolluted rural areas and may actually require a certain amount of sulphur in the air. Throughout Europe.

4

Lecidea limitata

Description: A crustose lichen forming thin grey sheets. These are commonly round or oval and often bounded by thin black borders. The spore-producing discs are round and black.
Habitat and range: Grows mainly on the trunks and branches of various trees, usually those species with smooth bark. It also grows on fences, but it cannot tolerate much pollution and occurs only in rural areas. Most of Europe.
Similar species: Several other lichens, including various *Graphis* species, form similar crusts, but their spore-bearing surfaces usually form straight or wriggly black streaks.

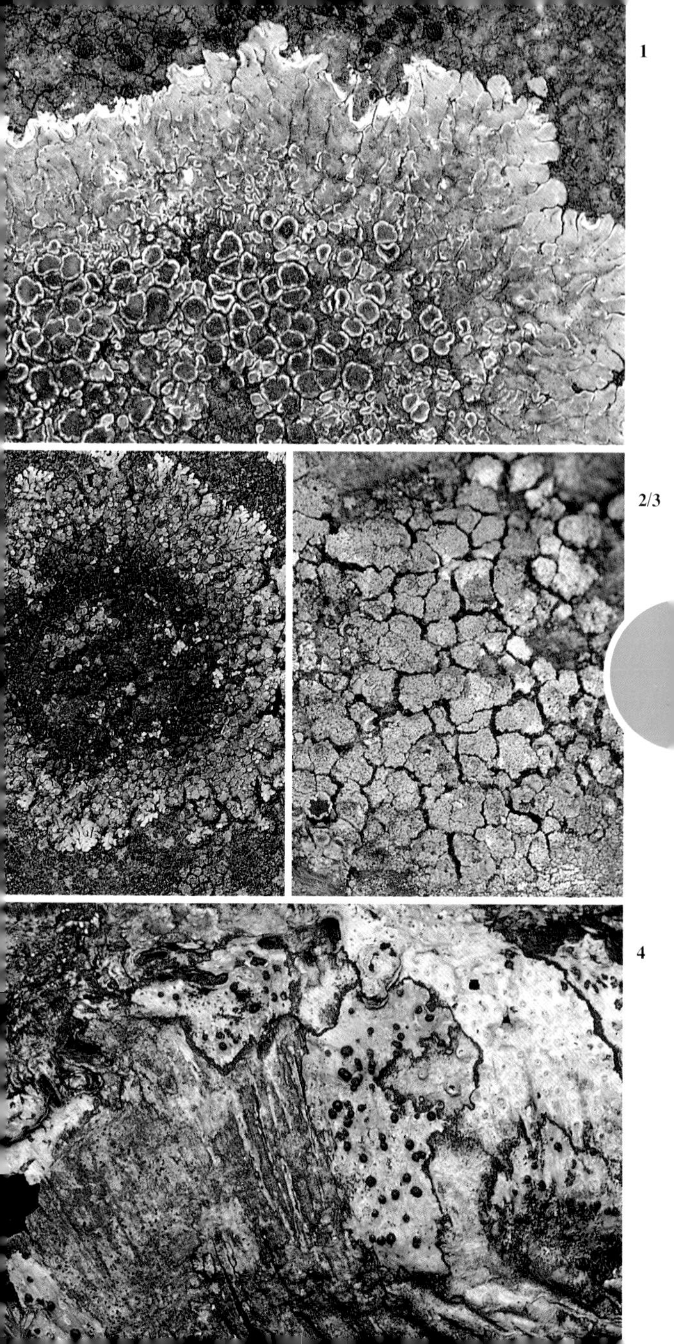
1
2/3
4

1

Parmelia sulcata

Description: A foliose lichen, whose grey-green thallus starts off more or less circular but usually becomes very irregular as it grows. It is composed of numerous fan-shaped lobes bearing a network of fine white lines. Clusters of powdery granules develop along these lines, especially near the ends of the lobes, and blow away to grow into new lichens (see p.246). True spores are rarely produced in this species.
Habitat and range: Grows on the bark of various trees; very common on old apple trees. Common in town parks and gardens, but in polluted areas it tends to occur only on the bases of the trunks – in common with *Hypogymnia physodes* (below) and other *Parmelia* species. The cleaner the air, the higher the lichen can be found on the trees. Most of Europe.
Similar species: Several other *Parmelia* species are superficially similar, but only *P. saxatilis* has the white lines, and this species lacks the powdery granules. See also *Hypogymnia physodes* (below).

2

Hypogymnia physodes

Description: A grey foliose lichen forming irregular patches of smooth and inflated, hollow lobes. The tips of the lobes split open to reveal masses of powdery granules. Spore-bearing discs are rust-coloured with grey margins; uncommon. The dark brown underside of the thallus is firmly attached to the substrate, without the anchoring threads found in *Parmelia* species.
Habitat and range: Abundant on tree trunks and wooden fences, including those in town parks and gardens, and often jostling for space with *Parmelia* species: also grows on old walls. Throughout Europe.

3

Cup Lichen

Cladonia fimbriata

Description: Intermediate between foliose and fruticose lichens, this species forms mats of greyish green scales which send up stalked cups, up to 4cm high and shaped like long-stemmed wine glasses. Scales and cups are all covered with powdery granules that can grow directly into new lichens when detached; cups also bear spore-producing discs on their rims.
Habitat and range: Abundant around the bases of old tree trunks and on dead stumps and old fence posts: also on rockery stones and bare ground, including lightly-used gravel paths. Throughout Europe.
Similar species: *C. pyxidata* is a very similar cup lichen but its cups are shorter and they widen more gradually towards the top – more like ice cream cornets than wine glasses.

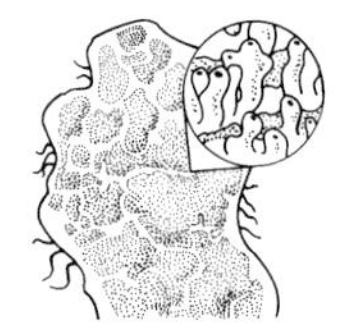

Parmelia saxatilis. Note finger-like 'snowmen' erupting from the lines

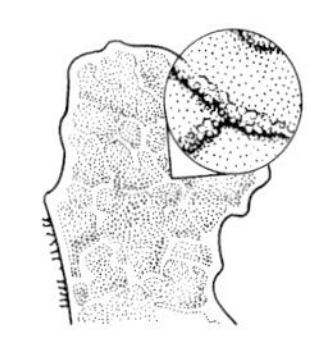

Parmelia sulcata. Note powdery clusters on the lines

1

2

3

Index

Main entries appear in **bold**.

Picture credits

The copyright in the photographs belongs to:

FLPA: M Silvestris: 23 b

Natural Image: Bob Gibbons: 137 b; 159 cl; 219 cl. Mike Lane: 27 tl; 47 b; 209 bl. RSK: 53 c. Peter Wilson: 61 c; 133 t; 145 c; 157 c; 161 b

Natural Science Photos: M Andera: 19 bl; 21 t, c, b; 193 c; 195 c. Alan Barnes: 31 c, b; 33 b; 37 tl; 39 b. Gillian Beckett: 213 c; 215 t; 219 t; 221 tr; 223 t; 225 tr; 231 c, b; 233 c, b; 239 b; 245 tl. S Bharaj: 69 t; 201 br; 217 c; 227 t; 233 tl. Christopher Blaney: 15 t, c, b; 37 tr. Derick Bonsall: 37 b; 39 t; 43 t, b; 45 c. PA Bowman: 53 b; 93 c; 105 tl; 161 c; 183 bl. Jeremy Burgess: 77 c. W Cane: 203 bl; 225 tl. Michael Chinery: 49 b; 51 c; 53 t; 55 b; 57 b; 59 c, b; 61 bl, br; 65 t, b; 73 c; 75 b; 81 cl, cr, br; 83 t; 85 cr; 89 tr; 97 t; 99 cl; 101 c, br; 103 tl, cl, bl, br; 105 cl, br; 111 c; 115 tl, tr, bl, br; 117 tr; 119 t; 121 cr; 123 t; 125 cr; 127 t, b; 131 br; 137 c; 139 b; 141 cr; 143 t, c; 147 b; 149 t, b; 151 c; 153 t, c; 155 c; 157 t, b; 165 b; 171 t; 173 c; 179 t, c; 181 t, b; 185 t, c; 189 tr, c; 197 t; 199 t; 201 t, c; 203 t, br; 205 t, cl, cr, b; 207 t, bl, br; 209 tl, tr; 211 tl, b; 215 bl, br; 221 bl; 223 c; 227 b; 229 t, b; 239 tl, tr; 241 c, b; 243 t, b; 247 c; 249 cr; 251 t. Steve Downer: 19 br; 41 b. Martin and Dorothy Grace: 219 cr; 225 br; 227 c; 237 cl; 243 c; 245 tr. JA Grant: 63 t; 133 b; 141 cl; 147 c; 169 b; 183 t. J Hobday: 93 t. Adrian Hoskins: 85 cl; 89 tl; 251 b. Iris Lane: 55 c; 63 bl, br; 79 cl; 83 c; 91 cr; 103 tr; 121 b; 129 c, b; 131 t; 135 t; 143 b; 145 t; 147 t; 149 c; 159 t, b; 163 c, b; 165 t, c; 187 t; 189 tl; 195 t. G Matthews: 211 tr. D Meredith: 23 c; 39 c; 41 c; 83 b; 107 br; 121 cl; 167 c; 189 bl. NKD Miller: 103 cr; 141 b. TA Moss: 65 cl; 67 b; 75 c; 123 cr; 133 cr; 161 t; 171 b; 173 b. JR Mountford: 25 c; 43 c. MW Powles: 19 c; 25 t; 27 b; 31 t; 35 t, b; 101 t. Richard Revels: 17 t, b; 19 t; 25 b; 27 tr; 29 t, b; 33 t, c; 41 t; 47 t, c; 49 c; 51 b; 57 c; 59 tl, tr; 61 t; 63 c; 65 cr; 67 t; 69 bl; 71 tl, c, bl; 77 tl, tr, b; 79 t, cr, b; 81 bl; 85 t, b; 87 tl, cr; 89 c, b; 91 t, cl, b; 93 br; 95 c, b; 97 cl, cr, b; 99 cl b; 105 cr; 109 br; 113 tr, c; 117 c; 119 bl, br; 123 b; 125 cl, b; 129 tl, tr; 131 c, bl; 133 cl; 135 c; 137 t; 139 c; 141 t; 145 b; 151 t, b; 155 t, b; 163 t; 167 t, b; 169 t; 175 t, b; 177 t, c, b; 179 b; 187 b; 191 t, b; 193 b; 195 b; 197 c, b; 199 c, b; 201 bl; 203 c; 209 br; 213 t, b; 215 c; 217 tl, tr, b; 221 tl; 223 bl, br; 225 bl; 229 c; 231 t; 233 tr; 235 t, c, b; 237 t, b; 241 t; 247 t; 251 c. OC Roura: 35 c; 45 b; 49 t; 57 t; 247 b; 249 t, cl, b. Roger Thomas: 45 t. PH Ward: 55 t; 69 c; 73 t; 87 tr, cl, b; 99 t; 101 bl; 105 tr, bl; 107 cl, bl; 109 tl, tr, c; 111 b; 113 tl, b; 117 tl, b; 119 c; 121 tl, tr; 135 b; 183 c, br. PH and SL Ward: 67 c; 69 br; 71 tr, br; 73 b; 75 t; 81 t; 93 bl; 107 tl, tr, cr; 109 bl; 111 t; 139 t; 153 b; 169 c; 171 c; 181 c. Andrew Watts: 125 t; 127 c. Ian West: 23 t; 29 c; 51 t; 95 t; 115 c; 219 b; 221 br; 237 cr; 245 b.

Nature Photographers: NA Callow: 173 t. Paul Sterry: 123 c; 185 b; 187 c; 189 br; 193 t

Wildlife Matters: 191 c